D0436010

A Gift
from
FRIENDS
OF THE
PALO ALTO
LIBRARY

SPARK

how creativity works

SPARK

Julie Burstein

foreword by Kurt Andersen

HARPER

An Imprint of HarperCollins*Publishers*

www.harpercollins.com

HarperCollins books may be purchased for educational, business, or sales promotional use. For information, please write: Special Markets Department, HarperCollins Publishers, 10 East 53rd Street, New York, NY 10022.

FIRST EDITION

Designed by Joy O'Meara

Library of Congress Cataloging-in-Publication Data

Burstein, Julie.

Spark : how creativity works / Julie Burstein ; foreword by Kurt Andersen. — 1st ed.

p. cm.

Includes bibliographical references.

ISBN 978-0-06-173231-7 (hardback)

1. Creative ability. 2. Creation (Literary, artistic, etc.) 3. Gifted persons—Interviews. I. Andersen, Kurt. II. Title.

BF408.B866 2011

153.3'5—dc22 2010028826

11 12 13 14 15 ID/RRD 10 9 8 7 6 5 4 3 2 1

For my mother, **Janet Handler Burstein**

Contents

Foreword

by Kurt Andersen
August 6, 2010

I graduated from college with no job in the offing and no desire to return home to Nebraska. All I knew for sure was that I wanted to live in New York City, hang out with people doing creative work, and get paid for doing creative work myself, but that I didn't know how to act or sing or dance or play an instrument or draw. When I was twenty-one, that was the extent of my career plan. And oddly enough, I've executed it in all its half-assed, unkempt glory for the last thirty-five years: I'm a New Yorker; my friends are mostly writers and artists and filmmakers and musicians and designers, and I've earned my living in pretty much every creative field that doesn't require me to make music or draw. Or dance.

But it was just a decade ago that I had two back-to-back aha moments that finally explained my zigzagging professional path to myself and also made me understand the prerequisites for creativity.

The first lightbulb went off when I read an essay called "The Amateur Spirit" by the great scholar and writer Daniel Boorstin. The main obstacle to progress is not ignorance, Boorstin wrote, but "pretensions to knowledge. . . . The amateur is not afraid to do something for the first time. . . . the rewards and refreshments of thought and the arts come from the courage to try something, all sorts of things, for the first time. . . . An enamored amateur need not be a genius to stay out of the ruts he has never been trained in."

Here was a supremely credentialed prince of the Establishment, the ultimate professional intellectual—Rhodes Scholar, Ph.D., professor at the University of Chicago and Cambridge University, museum director, Librarian of Congress—arguing in his seventies that while professionalism of the good kind (knowledge, competence, reliability) has its place, it is the curious, excited, slightly reckless passion of the amateur that we need to nurture in our professional lives, especially if we aspire to creativity in the work we do.

A few months later I found myself interviewing my funny, brilliant friend Tibor Kalman, the graphic designer and multifarious auteur. A transcript of our conversation would appear in a monograph about his work. He was forty-nine and when we talked he knew he had only months to live. Tibor had always been smart about the nature of creative work, but now the wisdom was pouring out.

"You don't want to do too many projects of a similar type," he told me. "I did two of a number of things. The first one, you fuck it up in an interesting way. The second one, you get it right. And then you're out of there. I have sought to move into as many other fields as possible, anything that could be a step away from 'graphic design,' just to keep from getting bored. As long as I don't *completely* know how to do something, I can do it well. And as soon as I have [completely] learned how to do something, I will do it less well, because what I do will become more obvious."

I realized my entire professional and creative life so far had been conducted in a similar way, by indulging the amateur spirit: I'd repeatedly, presumptuously barged into jobs for which I had no credentials or much specific training and then worked extra hard, hoping that my rank inexperience might somehow be transmuted into interesting innovation. I'd had no experience writing radio and TV news scripts (for NBC, my first job), or about politics or crime (for *Time*, my second job), or about architecture and design (also for *Time*), and when I cofounded *Spy* magazine (my fourth job), I had never edited anyone's writing but my own, or run a business. Ditto when I wrote and produced prime-time network comedy specials (for NBC), wrote an off-Broadway revue, wrote a screenplay (for Disney), and sold my first novel (to Random House). Professor Boorstin and my friend Tibor had convinced me retroactively that what I'd done by accident, going from interesting gig to interesting gig with no real strategy, had a philosophical basis.

Shortly after that double epiphany, executives from Public Radio International and WNYC called me out of the blue and asked if I might be interested in hosting a new program they wanted to create about the arts and entertainment and creativity. Really? *Me?* My total on-air experience consisted of having been interviewed a few times about books and articles I'd written. Host a weekly show on public radio? Were they serious? I'd done plenty of things I had no standing to do, but no one before had ever *invited* me to do something I had no standing to do.

That's not completely true. Twenty years earlier, a theater director had called me out of the blue and asked if I might be interested in playing the lead in his upcoming production of *Othello.* Really? *Me?* My total acting experience consisted of playing Captain Hook in a grade school production of *Peter Pan.* And also, I am, um, er, Caucasian. Was he serious? Well, as it turned out, um, er, uh, *no*: he'd meant to call an (African American) actor named Curt Anderson. Wrong number.

But this time, it turned out, the public radio grown-ups really had intended to call me, and not the veteran radio personalities Curtis Andreessen or Karl Andrews or Carter Andrazs. They were serious. And that's how I came to help invent and host *Studio 360.*

What we do every week on *Studio 360* is try to show how creativity works by means of individual case studies, by talking at length and in depth to some of the world's most talented people about how and why they do what they do. And for this book we've distilled the most relevant wisdom from my hundreds of conversations to create a kind of plain-English master class about the difficult, exhilarating process of pursuing one's creative passions. It's Creativity 101 featuring guest lectures by visual artists and designers Chuck Close, Denise Scott Brown, and Robert Venturi; filmmakers Kathryn Bigelow, Ang Lee, Mira Nair, and Kevin Bacon; writers Richard Ford, Joyce Carol Oates, John Irving, and Tony Kushner; musicians Patti LuPone, Rosanne Cash, Robert Plant, Yo-Yo Ma; and many other artests. Maybe you're an artist or would-be artist yourself; maybe you're an amateur singer or painter or writer. If so, consider this a collegial primer on how some supremely talented and successful people unleashed their talents and achieved their successes. But I'm also convinced that there are plenty of valuable, hard-won lessons about living and working creatively that can be applied to almost any life and any job. Or maybe you simply want to enjoy an unbuttoned, intimate look at the life and times of a few dozen cultural superstars. If so, enjoy.

What I've realized after talking to this remarkable pantheon of creative people for our five hundred shows is that what I learned from Daniel Boorstin and Tibor Kalman a decade ago is true of pretty much all work worth doing, especially creative work: the prerequisite for doing exciting work is to be excited about it yourself, reaching to do or make something that you haven't done or made before and which seems at least a little scary, just beyond your comfort zone. E. B. White famously wrote that "no one should come to New York to live unless he is willing to be lucky." The same goes for people who want to do any kind of creative work.

As soon as I adopted this paradigm of the amateur spirit just over a decade ago, taking risks to try new things, staying out of ruts, refusing to be paralyzed by the fear of imperfection or even failure, opening myself to luck—that is, once it became my *conscious* MO rather than simply the way I'd unthinkingly stumbled through life—I began spotting other members of the club, such as Danny Boyle, the director who made *127 Hours, Slumdog Millionaire, Trainspotting*, and eight other feature films. "Everything after the first one," he told the *New York Times*, "is business. There's something about that innocence and joy when you don't quite know what you're doing." And Steve Jobs, talking about the unexpected upside of being purged from Apple nine years after he founded the company. "The heaviness of being successful," he says, "was replaced by the lightness of being a beginner again, less sure about everything. It freed me to enter one of the most creative periods of my life." A period during which, among other things, he founded the amazing animation studio Pixar.

I'm not much of a religious person, but if forced to choose I'd probably go with Buddhism, because its practitioners write and say paradoxical things, such as this line by the Zen master Shunryu Suzuki: "In the beginner's mind there are many possibilities, but in the expert's mind there are few." That's what Tibor was getting at, and Boorstin and Boyle and Jobs. And Richard Serra, as he explained a few years ago in a conversation on *Studio 360*, which we've included in Chapter 2. "I'm just going to start playing around," Serra told me about his decision to abandon painting as a young man, "without the faintest idea of what I was doing."

I learned how to make a national radio show by making a national radio show in the company of people who knew lots more about radio than I did, especially Julie Burstein, my executive producer from 2000 through 2009. Having written for TV and radio and the movies, I knew how to write sentences for the voice and ear rather than the eye, and I knew how to tell stories. But I learned how to have a new kind of conversation, in which I uttered sentences that parsed and contained a minimum of *ums* and *uhs* and *you knows*, conversations in which I seldom interrupted but nevertheless took the lead.

Moreover, in creating *Studio 360* with Julie and the rest of our team of producers, I had the same goal as when I'd created magazines and websites and produced TV shows and written novels—to make a thing that *I* would want to read or see or hear even if I'd had nothing to do with it, and that was unlike anything extant. For me, that's also how creativity works, when it works. In this sense, creativity is selfish—but it derives from what I call "good selfishness," something like good cholesterol.

In the ten years that I've hosted the show, I've had more than a thousand conversations with some of the most creative and interesting people on earth. Many of them have surprised me. Before I met Susan Sontag, for instance, I was terrified. She'd been a hero of mine for decades, and her assistant had informed my producer that "Ms. Sontag does *not* suffer *fools*," just in case I happened to be one. But our hour-long talk turned out to be one of the best I've ever had—and the only one for *Studio 360* that generated a handwritten thank-you note. I was very differently surprised by the novelist and journalist Nick Tosches, who did his best to offend me and then, failing to do so, left the studio for a smoke halfway through the show and never returned. I was surprised when Gore Vidal remembered he had once threatened to sue me for an article I'd published about him, surprised when Twyla Tharp started crying, surprised when Rosanne Cash became a close friend, and surprised when Neil Gaiman asked me, years after he'd appeared on the show, if I would write a piece of short fiction for an anthology he was editing—and thus last summer I published my first science fiction story. Once again, I'd never done it, didn't know for sure if I could do it, but did it anyway, and was pleased with the result. Such is the terror and delight of trusting one's amateur spirit, being willing to be lucky and seeing where creativity takes you.

Introduction

When I was just starting out in radio, one of my first assignments was to interview the cellist Yo-Yo Ma and the pianist Emanuel Ax. I was barely out of college; the musicians weren't much older.

They were rehearsing at the Upper West Side apartment of Ma's in-laws for an upcoming concert in New York City. So I took the subway from WNYC's dingy studios in the Municipal Building and arrived at the apartment with my bulky Nagra open-reel tape machine. Ma met me at the door and ushered me into the living room, where I struggled to set up my audio equipment—not because I didn't know how, but because from the moment I walked in, Ma and Ax started teasing and joking with me, and I was laughing so hard it was difficult to focus on the thin brown magnetic tape as I wound it from reel to reel.

The entire interview progressed this way—cellist and pianist bantering and joking. I had been taught not to make a noise when the person I was interviewing was talking, never to say "Uh-huh" or "Oh!"—instead to just nod quietly so that my voice didn't interrupt. But Ax and Ma were so buoyant and hilarious that this was impossible, and finally I just gave in and laughed along with them.

Back at the station, I wrestled with the tape, trying to edit it for broadcast. This wasn't anything like what my conception of a "serious" music interview was supposed to sound like. Much later, I realized that the actual words the musicians had said were not the most important aspect of the conversation; I had been incredibly fortunate to experience the essence of their creative relationship. It was as if I was hearing the music behind the words, the feelings that gave their stories an emotional resonance. As Ax and Ma played together, their unrestrained conversation echoed the way they connected through music: as brilliant, talented, serious—and also mischievous—performers. Joy is a fundamental component of how their creativity works.

Throughout my career in public radio as a reporter, host, and producer, I've listened carefully for those moments when an artist, while telling a story, also reveals the underlying spirit of his or her creativity. When I designed *Studio 360* in 2000, I was determined to develop a show that would probe beneath the surface of contemporary art and pop culture to find the deeper currents that draw us in. It has been thrilling, over the first decade of the program, to listen to hundreds of artists and musicians and writers and filmmakers talk about where they find their inspiration, as well as about how they struggle with the process of making art, allowing us to see their vulnerabilities as well as celebrate their successes.

As I approached the new challenge of creating a book that would draw upon the hundreds of hours of conversations we have had in *Studio 360*, I began to think about some key questions: What do we look to art for, in the twenty-first century? What are these artists revealing to us, and why are we compelled to look and to listen? There are many answers to these questions; for me, the work that touches most deeply is always the work that connects with life. The artists with whom I fall in love are those who are willing to open themselves up to the anguish as well as to the pleasure of experience in order to create work that moves me to understand my own life in a new way.

So the deep threads I chose to follow as I arranged the chapters of this book can be found, as we said when we began *Studio 360*, "where art and real life collide." Perhaps even more aptly, they are where we experience the oscillation between art and life.

For *Spark*, I have selected stories from artists who tell us something about that oscillation, creators who turn to the people, places, and materials in their lives for their motivation and their subject. I've organized the book to explore emotionally resonant landmarks in both life and art. The chapters are a trail through challenges, triumphs, and transformations; they reveal connections to the natural world and to home and family; uncover the wonders of childhood and the frustrations and revelations of partnerships; and also touch upon disaster and its aftermath, when artists take the shattered fragments of the world and put them back together, for themselves as well as for us. Each chapter reflects a different facet of human experience, with which these artists wrestle and play.

In the final chapter, artists talk about how they get to work, with stories about productive beginnings, false starts, the need to step away sometimes, and how to recognize when something is actually finished. This is a crucial piece of the story: if they never got to work, there would be no movies, no poetry, no paintings, no music—no connection between these creators and us.

A few years after my conversation with Yo-Yo Ma and Emanuel Ax, I was asked to produce a national radio series for the Mostly Mozart Festival. The English playwright Peter Shaffer was in town, and I wrangled an interview with him for the radio show because his play *Amadeus* had just been made into an Oscar-winning movie.

Shaffer explores the relationship between Wolfgang Amadeus Mozart, whose genius was brilliant and unpredictable and enduring, and Antonio Salieri, a court composer who had a long, celebrated, and ultimately pedestrian career. When I asked Shaffer what he saw as the central difference between the two men, he told me that he pictured Mozart running up to the well of inspiration and diving in, headfirst, without stopping, while Salieri walked up to the side of the well and peered over, wanting to see what was at the bottom before he dipped his net.

The artists you'll meet in *Spark* are all willing to dive, head-first, into the raw, exhilarating, and sometimes unfriendly experiences of life. Listening to them talk, hearing their stories about their methods and their catalysts, doesn't diminish the power of their art: it deepens our appreciation for it. Their stories also tell us something about how to open up to allow creativity into our own lives.

SPARK

Chapter 1

ENGAGING ADVERSITY

In the spring of 2000, when I was brought in to lead the team that would launch *Studio 360*, the public radio landscape was already littered with the dried husks of failed programs about art and culture. During the interview process for the job, I had outlined for Public Radio International (PRI) and WNYC my vision of a show that looked at life and ideas through the lens of creativity, pop culture, and the arts: to explore, as we later said in *Studio 360*'s tagline, the place "where art and real life collide." Once I was hired, I faced some internal challenges as well: the team of talented producers had been demoralized by a difficult development period prior to my arrival; we had just three months to develop a new idea and launch a national weekly series; and our promising talent, the novelist and journalist Kurt Andersen, had never hosted a radio show.

I knew from my first few conversations with him that Kurt had the capacity to become a great radio host. But I had been charged with creating a new magazine-style show, and in public radio that means the host is required to read pages and pages of intros and outros for produced stories. This requires a lot of practice to do well, and at the beginning, reading copy is often frustrating.

Kurt worked hard and learned quickly; listening to our first show again now, I'm still captivated, for example, by his initial commentary on the blurring between advertising and content on television, which was sharp and funny, and which he delivered with verve. But I knew that if reading copy was all Kurt did, listeners would never really get to know him.

So I decided to throw out tradition and explore an idea I'd played with in a couple of projects I'd developed before coming to *Studio 360.* I thought: What if Kurt wasn't alone in the studio reading intros? What if we chose a central idea for each show, produced stories that approached the idea from different perspectives, and then brought in smart, interesting guests to listen to stories with Kurt, and talk together about what they heard? Then too, given the range of subjects we wanted to explore, why not take Kurt out of the traditional objective narrator role and allow him to offer his own opinions in conversation with a guest each week?

This novel approach engaged the team and, it turned out, helped unleash their creativity. Each week, we chose a "cover story" for the show, gathered provocative stories from pop culture and the arts around this central theme, and found an accomplished artist or writer or actor or director to join Kurt in the studio and engage in an energetic conversation about the ideas the stories raised. Some of the cover stories were directly connected to the world of art and culture, others more far afield. In our first couple of years, Kurt spoke with Nora Ephron about cooking, and with Susan Sontag about war; with Rosanne Cash about creative children of famous parents and with Simon Schama about the way maps help us understand the world. We loved that this new form allowed us to control many aspects of the production, and that it also left room for surprise and spontaneity. The conversations revealed unexpected facets of the famous guests: Who knew that the architect Frank Gehry loved the novels of George Eliot and drew inspiration from them as he developed new ways to frame views in his museum projects? Or that Madeleine Albright believed that "jazz and those who play it are America's greatest ambassadors as a reflection of our democracy"?

Listeners were intrigued. At a public radio convention a year or so after the show launched, the veteran NPR host Susan Stamberg told her audience that *Studio 360* was her new favorite show. "I love the format," Stamberg said, "the idea of Kurt Andersen and his radio pal of the day—somebody who's really smart—and together they listen. I love the notion of a theme: reflections of the society and the culture, presented in such interesting ways! I always come upon new ideas and a new way of thinking about things, and it's such good company for that hour—I'm just crazy for that program!" *Studio 360* became the fastest-growing show in Public Radio International's history.

Kurt is now a seasoned host, as comfortable taking on a traditional host's role as with pushing its boundaries. *Studio 360* currently employs a more familiar magazine format of interviews, produced stories, and host narration, as well as, at times, venturing into more complex documentary work through the show's American Icons series. Yet I still love to listen to programs from those early years, which were so ambitious and possess a liveliness and spontaneity that is hard to achieve in a highly produced show. Had Kurt been an experienced radio talent when we began *Studio 360*, I might never have gambled on taking a new approach.

Engaging challenges often force us to create stronger work. The adversity may be intensely personal: painter Chuck Close reveals the serious, undiagnosed learning disabilities he struggled with as a child, and how they shaped both his view of the world and influenced the iconic portraits that he creates; poet Donald Hall describes the year he spent caring for his wife, the poet Jane Kenyon, through a devastating illness, an experience that spurred him to write some of the most emotionally resonant work of his long career.

Even setbacks that aren't so dire still have the capacity to derail creativity. Money troubles plague many artists: singer and songwriter Jill Sobule talks about how financial backing—or the lack of it—forced her to improvise a new way of producing an album.

Although these artists face crises, their message isn't bleak. Instead, their stories illuminate how they use imagination, humor, and the grounding nature of their art to confront and survive difficulties.

Painter Chuck Close

Breaking Down a Face

"In the forties or fifties nobody knew from learning disabilities; I was just dumb."

In 1967, Chuck Close took a series of photographs of himself with his eyes half open behind dusty black eyeglasses, hair wispy and wild around his head, a cigarette dangling from his lips, and smoke curling up one of his nostrils. He chose one from among the many images and drew a grid of lines over it, twenty-one across and twenty-six down, breaking down the photograph into small squares. He then took that grid and exploded it onto a rectangular canvas almost nine feet high and seven feet wide. Into each six-inch square he airbrushed tiny black dots, then erased all evidence of the penciled network of lines. Through this painstaking process, Close created his startlingly detailed, and aptly titled, *Big Self-Portrait.*

Portraits were considered hopelessly old-fashioned at that time in the art world. "When I first started painting, not only was painting dead, but representational painting was even more so," Chuck Close said. "And the most bankrupt of all genres was the portrait. In fact, the great critic of the time was Clement Greenberg who said, 'The only thing an artist can't do today [is] paint a portrait.' And I thought, 'Hmmm, that means I won't have much competition. I'll see if I can breathe new life into something which seems totally derelict.'" Close quickly achieved that goal: his paintings from the 1960s onward, of such friends as Richard Serra and Philip Glass, captivated the imagination of viewers and critics alike.

Since then, faces have remained his focus. He chooses to paint people he knows, capturing friends and family in photographs that he uses as the basis for huge portraits. Like his initial self-portrait, these are not glamorous images; Close himself often refers to them as "mug shots."

Over the years, he has played with various materials and ways of making marks, defining faces with sprayers or brushes, oil paints or pastels, fingerprints or bits of colored paper, silken threads or woodblock prints. But his process of breaking down an image into parts has stayed constant. This method, he said, grew directly out of trying as a child to figure out the world. "I was learning disabled, although in the forties or fifties no one knew from learning disabilities; I was just dumb. I couldn't memorize; I still don't know the multiplication tables. It's sort of ironic that I use a grid and things that seem mathematical. They're all found and felt; I don't do any measurements with a ruler. I work it all out without the use of a calculator."

When he was five years old, Chuck Close asked his father to make him an easel for Christmas. "Shortly thereafter I nudged [my parents] to get a box of genuine artist's oil paints from the Sears catalog, so I think I was off and running at an early age. I was studying privately with a woman and drawing and painting from a nude model at age eight. Which made me the envy of everyone in my neighborhood."

Born in Monroe, Washington, Close spent his childhood moving around the state. His father was a metalworker and inventor who died when Close was eleven, the year his mother, a pianist, was diagnosed with breast cancer. Art was a refuge as Close and his mother moved from town to town. "I learned early on that since I wasn't athletic—I couldn't run or throw or catch a ball—and I was an only child, I needed to do something to keep people around me. So I began doing magic acts and puppet shows and things to entertain the troops. And I think that carries through; I began to realize that one of the things I could do that my friends couldn't do was draw. I think that everyone needs to feel special. I was lucky enough to have the support of my family and to feel like I had something to say even though I didn't say it in the traditional way, or couldn't spit back facts and figures and names and dates."

An eighth-grade teacher told him to set his sights low and consider vocational school because of his disabilities. But after studying at a community college, Close enrolled at the University of Washington. On the strength of his artwork he went to Yale for his MFA, then traveled through Europe on a Fulbright.

His early work looked nothing like the *Big Self-Portrait* that launched his career. "The way I arrived at what I was doing as a mature artist is almost completely a reaction to what I had been doing as a student. My work had been abstract, loose, sloppy, and I worked all over, all at once. I didn't know when I started or finished. I wasn't very good at flying by the seat of my pants.

"So I wanted to have something very specific to do every day, and I wanted to do today what I did yesterday, and tomorrow I'll do what I did today. I didn't want to reinvent the wheel every day. I found a way to break down a complicated image into a lot of small bite-size pieces. Also, part of my learning disability was being overwhelmed by the whole, and I found it to be particularly helpful to use a grid to isolate one small piece that I could work on and forget about the rest of the picture. All of my work since the mid-sixties, even those early continuous-tone black-and-white paintings, which seem seamless and graphic, were actually made, even then, in incremental units. Large, six-inch-square chunks that would later fit together seamlessly, so no one was aware that they were built incrementally. But indeed they were."

He's also certain that his choice of subject stems from another facet of his learning disability. "I've had face blindness, or prosopagnosia, my whole life. Really, it's been a nightmare situation for me," Close once said. "I don't know who anyone is and have essentially no memory at all for people in real space, but when I flatten them out in a photograph, I can commit that image to memory in a way; I have almost a kind of photographic memory for flat stuff." Photographing and painting the portraits of people he knows and loves gives him a chance to remember their faces.

Breaking down an image into a grid is nothing new; artists have used the technique for centuries. "I think it goes back certainly to the Renaissance, but I also think it went back to the ancient Egyptians," Close said. "But certainly from the Renaissance onward, the grid was a scaling device, a way to enlarge something. You see, in drawings of Michelangelo, often a preparatory sketch will have a grid transposed on it. That's used as a way to scale something up. This has been used forever, and it's something that is a convenience. At first, I used it just to locate where I was and concentrate in one area. But at a certain point in my work I began to let the grid show, to leave it as part of the picture. From that point on, the work has had its incremental nature [as] part of the structure and way to build the painting, but also [as] a self-evident way to understand it."

The painter loves that the grid gives him a way to play with one of the inherent contradictions in making paintings: the ability to create, on a flat surface, an image that suddenly appears three-dimensional. "It's also a record of the thought process that the viewer can get in sync with. I was recently living in Rome, and I was looking at the humble Roman floor mosaics. And the viewing distance is the height of the viewer; you look down on the floor made up of chunks of stone. Just when you get used to looking at the chunks of stone on the flat surface, it warps into an animal or a head or whatever. And just when you're comfortable looking at the animal or head, it flattens back out and insists on being seen as chunks of stone on the floor. The wonderful thing is that across the centuries, it's as if I'm looking over the shoulder of the artisan who made that floor. I can see the record of the thought process that he used where he chipped a little corner off the stone and nudged it in, and three or four or five or six of those stones become a piece of a face or an eyeball. It's something that the viewer can instantly understand as a process."

Close plays with this technique in all of his paintings, toying with the viewer's perceptions. "The flat surface is artificial; it warps into the real, and it goes back into the artificial. Which is one of the reasons I had trouble with the term 'realist,' because I was as interested in the artificial as I was in the real."

With his initial portraits, where he erased the grid, Close was considered a photorealist. Then, in portraits such as in *Robert/104,072*, which he painted in 1973–74, he began to allow the grid to show. That huge painting depicts a man with a mustache, wearing aviator glasses and an argyle vest. Robert is a junior high school friend of Close's wife, and Close portrays him by painting thousands of dots with an airbrush. The tiny grid can be seen if a viewer comes in really, really close to the canvas. "The first dot pieces I did, the dots were a sixth of an inch. So for a nine-foot-high painting there'd be about 104,000 dot-squares. Each square had one single dot sprayed in the middle of it."

Close often revisits old photographs, creating new work by changing his painting process. He's created several different portraits from his 1969 photograph of the composer Philip Glass with an airbrush, oil paints, or tiny balls of paper. "I've accepted other things rather than a complete, strict left to right, top to bottom grid. I did work with my fingerprints, and paper discs that are wafer thin, and those are still incremental but they don't follow a strict grid. And many people seem to find those works, which are not fastened to a grid per se, but are still built out of units, to be particularly emotionally laden. . . . Robert Storr, the curator of my retrospective at the [Museum of] Modern [Art], talked about my fingerprint paintings as being almost like caressing the face of the person that I'm painting. And the idea that skin makes skin was a nice idea, and also makes art that can't be forged."

In the late 1980s, Close began experimenting with other changes, placing his giant portraits on dark backgrounds so that the faces seem luminous by contrast. In 1986 and '87, he created a series of portraits of the artist Lucas Samaras, whose brooding eyes stare out from a pale face surrounded by a penumbra of brown hair and black background. In these paintings, sometimes the grids are circular, and they're visible, and larger than in the past, giving the artist "more room inside each square. I could put more than one simple dot; I could have two or three or four colors, then the grids got bigger and coarser yet. As they got bigger they got more complicated. They would run together. They could be hot dog shapes or doughnuts or sock shapes. So I just found them in the act of painting." If you closely examine a corner of one portrait of Samaras, you'll find that a patch of his forehead is made up of hundreds of small bull's-eye paintings in pale pink, blue, and gold, which, as you step back from the canvas, blur together to form a bit of skin on his face.

This new direction came just before a great trauma in Close's life: in 1988 he was struck by a very rare spinal aneurysm and within hours was paralyzed from the neck down. Close was forty-eight years old—the same age his father had been when he died.

It took months of rehabilitation at the Rusk Institute in New York City before Close could sit in a wheelchair, but he was determined to paint as soon as he was able. First, he held the brush in his teeth, and then, as he gained some movement in his arms, he strapped the brush to his wrist and held it steady with both forearms. Years later, his friend the painter Mark Greenwold said that as Close became stronger, "work became such a joy" for the artist. "Something in the act was affirming. Did art save his life? I think so."

Since what Close calls "The Event," he has remained in a wheelchair, and daily activities such as eating and traveling have been completely altered. "There are things I miss, walking on the beach, swimming, playing catch with my kids," he said. Painting, however, is "the aspect of my life in which I'm most the way I used to be." Close had always painted sitting down, and he said that when he paints now, "I'm sitting in a chair and I'm looking at a world that's essentially unchanged. Then sometimes I roll by a mirror and I'm startled to see myself as other people see me."

The major change in his work has been a move from the somber black-and-white of his early career to the exuberant color in his post-Event portraits. Up close, a canvas looks like a pixilated riot of squares filled with brightly colored circles and doughnuts and hot dogs of pigment, warped lozenges of golden yellow and lime green and fuchsia. As a viewer pulls back, he or she will find the perfect distance where the magic happens again, the colored shapes blend into eyes; a nose emerges; and suddenly there is a face.

In his later work, Close also plays with the orientation of the grid, sometimes turning it forty-five degrees so that each box is now a diamond. "Often I do that because the insistent beat of the grid, when it's horizontal or vertical, something about the way our brain functions sees that grid in a more insistent way. You can never quite get away from the grid, but by tilting it at an angle, our brain doesn't process it in the same way, so it loses some of its insistent rhythmic beat. Formally it gives a pinking-shears edge to the outside of the form as it zigzags down the canvas."

This is true of his 1997 self-portrait, which exhibits another difference as well. Although most of his paintings capture the subject head-on, in an unforgiving pose like a driver's license photo, in this one, Close is turning slightly over his left shoulder to gaze at us. He has cropped the photographic image so that in the painting his calm face fills the entire eight-and-a-half-by-seven-foot frame. We can see that this is the same man who glared down into the camera thirty years before, but here Close is relaxed, bald and elegant with a graying goatee and intense blue eyes behind round black glasses.

"I know my paintings are going to look a lot like the photo I'm working from. I graphically will be ahead. I know where I'm going to end up but I don't know the route I'm going to take. So much is embedded in the process of following that path wherever it leads, and the things you bump into, the ideas that occur to you through the act of painting, through the process of building a painting, are so different from the ones that you sit around and dream up. I don't wait for inspiration. If you wait for the clouds to part and be struck in the head with a bolt of lightning, you're likely to be waiting the rest of your life. But if you simply get going something will occur to you."

When Close appeared on *Studio 360*, he said he had begun to investigate a new technique. "I'm doing pieces now that are scribbled with no grid at all. Scribbling twenty different colors. Every once in a while I feel the need to get outside the box—no pun intended—but there is something that always seems to draw me back to it as a working methodology, and I think it's that I found so much elbow room within it. I found I can function more intuitively. When I could make any shape I wanted on any part of the canvas, I'd make the same four, five, six shapes over and over. When I could use any color I wanted I'd use the same four or five color combos over and over. Once I accepted the limitations of working within the incremental grid, I found myself making shapes I'd never made before. Using color I'd never used before. So it's always seemed liberating rather than constricting."

Close had also begun a series of portraits employing the early photographic technique of the daguerreotype, which portray each sitter's image in silvery grays. "The interesting thing about daguerreotypes is that it was the first form of capturing an image. There was camera obscura: people found an uneven way with a pinhole to get a projected image. But it was elusive. It disappeared. And everyone wanted to find a way to fix that image. The way to fix that image [came] in the 1840s, [when Daguerre] found he could get a polished sheet of silver that either absorbed or reflected light. For me, they're very elusive images. Only one person can look at it at a time."

Close has taken daguerreotype images of himself, of course, as well as of friends and celebrities such as the actor Brad Pitt and the model Kate Moss. He's used those images to create silk tapestries eight feet tall, in which each face emerges out of inky blackness, as if floating up from a dark pool. There is no penciled graph, but the warp and weft of the threads create one, harkening back to an early influence on the artist. "I know that one of the important primal experiences for me as a child was watching my grandmother knit and crochet and make quilts and afghans and things like that, which look a lot like my work today. She would crochet pieces and put them together to make even bigger pieces. A lot of what I do has a lot to do with what was called women's work—a process that you sign on to and you keep working at it until you get something. I think it has a lot to do with construction, and I try to build a painting rather than paint it."

Close is fond of saying "Inspiration is for amateurs, and the rest of us just show up and get to work. But so much of it comes out of the process . . . if you try to preconceive everything you do and conceptualize it, you're gonna do the same thing over and over. If, however, you just get busy and things occur to you in the process, you make the rules and therefore you can break them."

Poet Donald Hall

Mourning on the Page

"I needed to put down my own grief."

Donald Hall understands the power of getting to work: writing poetry sustained him through perhaps the most difficult chapter in his life. In January 1994, his wife, the poet Jane Kenyon, was feeling under the weather, achy, like she had a cold that would not go away. Hall was out of town, giving a poetry reading, when Kenyon called him to say that she had a terrible nosebleed and that a friend was taking her to the emergency room. Hall immediately got on a flight to head for their home in New Hampshire. As the plane took off a thought came into his head: "Leukemia."

When Hall arrived at the hospital, where his wife had been admitted, he knew he was right, even before her doctor said a word. Jane Kenyon died fifteen months later.

Years earlier, Kenyon had written poems about death and mourning when she believed that Hall was going to die. "I was nineteen years older than Jane, and I had cancer a couple times while she was healthy," Donald Hall said. "In fact, only three years before Jane died, I had two-thirds of my liver out with metastatic cancer. I was not supposed to live. That is to say, the statistics were not encouraging. We both assumed I would die soon. She wrote beautiful poems in the expectation I would die. She in a way wrote memorial poems about me. There's such an irony to that. It's an irony that was there since the beginning, and it's obvious, and I live with it every day. It's there in the poems. I think of how it should have been, that she should be living and writing the poems about me. It's just further outrage. Further outrage."

Donald Hall spoke with Kurt for our show about memorials on the first anniversary of 9/11. During their conversation, Hall said that work kept him going during the months after Jane Kenyon died in the spring of 1995, when he began writing poems that sear the heart. "I needed to put down my own grief. And one whole book called *Without* was about her illness and death, and then the process of the first year of mourning. The more recent book *The Painted Bed* carries mourning further by necessity as it is further away from her death. But it is necessary for me to deal with my own feelings, to put them out there, to try to preserve them, to try to preserve her in some of her particulars, things that I remember with a special delight and tenderness. To preserve them for others. To preserve also the position of mourning, the state of grief. It seems important to keep this. Not to let it go. I think it can be important to people who are in a state of grieving or mourning to find it enacted by someone else."

In *Without*, a slim volume that Hall published in 1998, his poems are full of the universal terror and boredom of life in hospitals, with catheters and chemotherapy and long hours reading the newspaper by a loved one's bedside. But the poems brilliantly illuminate the particulars of *this* patient and *this* caretaker, both of whom are poets.

In the first poem, an unnamed man watches his wife as she lies in a hospital bed, they weep and whisper of their love for each other, a scene familiar in sickrooms anywhere. But then, when it snows, "They pushed the IV pump / which she called Igor / slowly past the nurses' pods, as far / as the outside door / so that she could smell the snowy air." That description immediately reveals a woman who loves playing with words so much she names her IV pump, and who is so passionately connected to the natural world that she must smell the snow. As well as a husband who does not say "Oh, it's too much trouble, you'll get too tired" but knows that wheeling her to the door is essential, then returns her to her hospital bed, sits down again in the uncomfortable chair next to her, and writes a poem. "I didn't begin the mourning poems until a few weeks after her death, but while she was ill, sitting beside her in the hospital, when we were hoping she'd live, I wrote poems about her illness. I read them aloud to her some. We were not a euphemistic pair. We faced things very squarely. And people may even call us a bit morbid, I'm afraid . . . We were all for calling things by their names, and out loud, not suppressing anything."

Hall continues this honesty in his poems, which describe in intimate detail the awful particulars of the disease, the infusions of chemicals, the total body irradiation that prepares his wife for a bone marrow transplant. And he captures the moments of life and liveliness amid the despair, as when he dons a paper surgical gown, hat, mask, and gloves, in order to visit his wife in her isolation, and Kenyon tells him he "looked like a huge condom."

The specificity, the descriptions of surroundings, the bits of remembered conversations are essential for the poet. "I think in reality what I'm seeking—and, when it works, what I am achieving—is the response to an event, but it has to have a particular clothing," Hall said. "It has to have a texture and a color, or story. In order to present the feeling that it started from, it cannot do it directly, it has to do it through story, through the image, through the argument, whatever. Finally, in the act of revising, in the act of achieving the work of art, you are detaching the reality of the description from anything that needs verisimilitude; you are in a way detaching feeling, but you are doing it by means of the necessary furniture of language."

After his wife's death, writing was his main comfort. "In *Without*, at one point, I talk to my dog before I sit down in the morning to work on my poems, and I say to him in a manic glee, 'Poetryman is suiting up!' And that's how I felt. For that first year after Janey died, I could work on the poems for a couple of hours perhaps, and then I had twenty-two miserable hours to wait until I could get back to them, because I can't work on poetry all day. But the two hours were the only hours of happiness at that time."

Hall said he never worried about whether it was appropriate to capture the personal tragedy of his wife's illness and death in words. "When I was a young poet—a would-be poet—I remember breaking up with a girlfriend, and on my way home from the terrible breakup, with lots of tears, a line of poetry came into my head, and I felt bad, I felt guilty that I was exploiting the suffering by making a poem out of it. I went home and worked on the poem. But I have no feelings of exploiting Jane's death. It's such an absolute necessity. Writing poems served me by bringing me closer to her in the moments of working on them and working them over. I work endlessly on poems, I have to revise everything a hundred times. And I did also look to her for help. She was a very good poet. Well-known these days, goodness! I constantly thought, 'What would Jane say here?' So there was a companionship in writing them as well. The poems perhaps brought me closer to Jane alive than I would've felt otherwise. Many of the poems in *Without* are addressed to her, they are addressed to 'You,' they are letters to her, written after her death. I did, like many widowers and widows, talk to gravestones, talk to photographs on the walls, but I also sat down and—using the art that I've been trying to practice for fifty years—addressed her with my art."

Hall's poems to his lost wife are full of local news as well as heartbroken lament. In one long poetic missive, "Letter with No Address," Hall asks Jane questions about what happens after death, describes his fragile state attending his small granddaughter's birthday party, "a week after you died, / as three year olds bounced / uproarious on a mattress" and his new routine of solitary meals and solitary walks with their dog, Gus, to take irises to his wife's grave. He writes about how he talks constantly to her photographs about their shared life: "Ordinary days were best / when we worked over poems / in our separate rooms." Hall ends by revealing that he drove to the graveyard three times that day, coming back home and imagining that "you've returned / before me, bags of groceries upright / in the back of the Saab / its trunk lid delicately raised / as if proposing an encounter / dog fashion, with the Honda."

"When I wrote *Without* . . . in the letters I wrote to Jane, just about all of them have something in it that's supposed to make her laugh," Hall said. "I don't tell jokes like the minister, the rabbi, and the Christian Science practitioner, but I say things that would in fact have made her laugh and would make others laugh. Some people, my goodness, are really upset with that. There are a couple of sexual jokes in the letters to her, and one English editor who'd published me before told me that he admired the poem up until the very end, but 'Donald, Donald—you couldn't do that.' Well, I did it. And Jane would've roared with laughter, and that's whom I am addressing somehow in my heart when I write these poems."

Reports on what is happening in the natural world of their old New Hampshire farm infuse these poems of mourning with the changing seasons. Hall tells his wife about the birds at their birdfeeder outside the kitchen window, about how he gazes beyond them to Mount Kearsarge and remembers her gazing out the same window, about the weather tending toward snow. In "Day-lilies on the Hill 1975–1989," which refers to the years Hall and Kenyon lived on the farm his family had inhabited for more than a century, Hall remembers further back to what it was like to trim the hayfields as a child with his grandfather, Wesley, who "works with the presence and practice of sixty years. / I watch him twist his fork in, balance, heave, balance, / and swing it over his head: so, so . . ."

In his long career as a writer, Hall has often observed the rhythms of the farm life into which he was born in 1928. "I grew up attached to the landscape I live in now, of New Hampshire. Which is a landscape where the whole of houses and farms were disappearing. When you walked in the woods you'd have to look out for cellar holes you might be falling into, or old wells. I grew up very aware of losses. And it wasn't sad. I celebrated this place as it had been. . . . I saw it disappearing, and I wanted to keep it as much as I could. The world I inhabited as a child in the 1940s in New Hampshire, haying with my grandfather in the summer, was an archaic world. No tractors, no bailers, one horse, a hayrack. I've written again and again about that kind of farming and what it was like to live there. To live in an economy that the amount of money a year would sound like super poverty, but nobody thought they were poor. They could burn wood for heat, they could take ice out of the pond for cool in the summer, they could shoot a deer a year, grow big gardens. It was a world without cash. I certainly don't praise that world, I don't want to bring it back. If we could bring it back I'm not sure I'd want to live there. I want to keep that world in the world by writing about it. . . . This is a motive to literature—preserving what is gone or what is going. And it is, of course, an important part of preservation to try to preserve the dead whom you loved and admired."

In one of the poems in *The Painted Bed*, Hall expresses the complicated idea of preserving a particular memory of his beloved wife. It's titled "Her Garden," and you can listen to the poet reading it on *Studio 360*. It begins:

I let her garden go.
let it go, let it go
How can I watch the hummingbird
Hover to sip
With its beak's tip
The purple bee balm—whirring as we heard
It years ago?

In the poem, Hall reveals the decay of Kenyon's beloved garden, slowly reclaimed by weeds and forest after her death. "She died in April, the flowers began to come up, still come up in diminished form. There are things of hers I keep up, that I preserve not just in poems. And there are things that I somehow can't, I don't want to. There's a certain, oh, bitterness in it. I can't imagine working over the garden to pretend it's the same as it was when she was here. For some reason it suits me to see them diminish. I love her flowers, they still remain. The peonies that still come up. Hollyhocks—not so good this year. But I don't want to restore that which was there before, except in a poem perhaps."

Hall's reading voice, strong and sonorous, infuses his poetry with feeling. He often travels around the country to give readings, and frequently chooses these poems of illness and loss for his appearances. "Again and again, people will say to me in a question period or a reception afterwards, 'How can you read them aloud?' I have no problem, I have no difficulty reading them aloud. They have become poems. They start out as screams of pain, and the first drafts are terrible. But all my first drafts always are, whatever the subject. I work over them and over them and over them, and

I love working over them. I don't write, I rewrite. As I work on them they will come to be something outside of me.

"It's as if I were hacking away the piece of stone to make sculpture, or modeling, or trying to paint, adding a dash here and there. So that they become works of art. They become a work of art in the art that I love and I've tried to practice since I was twelve years old. Then, they are then objectified. They start from the pain, they start from the anguish, these poems do, and then they become something that needs to be in itself as language, as sound, as imagery, objects of pleasure, objects that give pleasure. So they have—and this is true of so many poems—one of the paradoxes of poetry. I'm talking about poetry for four thousand years, the subject is so often defeat, death, and loss. Even with triumph, there are aspects of loss always, and yet the material itself is beautiful and gives pleasure. The process of working out of the raw material of this grief and loss and the raw material of the scream is to try to make it into something that does not change the death that is spoken of, does not alter things. But it does use that material in a way that is in itself to be an object of beauty."

Jane Kenyon's last request to her husband was that he be with her when she died, and in poems Donald Hall observes her last breaths and how he reaches over to close her eyes. He then holds out his broken heart to us in verses that convey the anguish of one who is left behind, as well as the inexorable pull back into a world filled with flowers and birds and family and even the possibility of new love.

Singer and songwriter Jill Sobule

Seeking Support

"It could have been humiliating;
the surprising part was how great everyone was."

At the TED (Technology Entertainment and Design) conference in 2006, Jill Sobule sang a sweet song about dogwoods in flower and people enjoying picnics on a lovely, warm day in Central Park—in January—and received a standing ovation led by none other than Al Gore. Over the years, Sobule has unleashed her satirical wit onto everything from global warming to what it takes to be a supermodel. Her music is part autobiography, part fantastic imagination; in a single song she can move from love to apocalypse, which won't be so bad, as she notes in "A Good Life," because when the world ends at least we won't have to pay our bills or make our beds.

As a kid growing up in Denver, Colorado, Sobule knew that rock and roll was her future. At six she was playing the drums; by middle school she had picked up the guitar. "I was always a strange girl, I didn't want Barbies, I wanted a Gibson. I wanted a Les Paul," Sobule said. "I had a brother seven years older than me who played guitar. And his guitar teacher was Chet Atkins's brother . . . So my brother was a serious player; there were guitars all around. And my brother was my hero back then, and my role model." He listened to John Prine and Randy Newman, and his little sister listened closely alongside. "I always liked how they would tell stories. And they'd always have a sense of humor but there would be something tragic at the same time."

Thirty years later, the same could be said about Sobule's witty, topical story-songs. Listening to her biggest hit from the 1990s, "I Kissed a Girl"—the first one, more matter-of-fact and definitely catchier than the vampy 2008 Katy Perry song—Sobule still sounds like someone's wiseass kid sister. But she's grown-up now and sitting next to you on a bar stool telling you funny stories about her life through clever pop vignettes.

While a couple of her songs, including "I Kissed a Girl" and "Supermodel" (featured in the film *Clueless*), have been hits, Sobule has had a complicated relationship with the record industry. To begin with, Sobule has said she's never made money from her albums. "I have never made a cent off a record in my life. I have never recouped enough, and I never sold enough. When people see you have a song on MTV, they think you are doing well—but you know, the way the traditional label deal was set up, it is really hard for an artist, unless they sold a lot, to see anything." After two major labels dropped her, she signed with "two indie labels that went bankrupt, [so] the thought of maybe getting another deal seemed ridiculous."

In 2008 Sobule captured her mordant mood with the song "Nothing to Prove," which begins with a description of her attempt "to impress someone at a dying record company." In the chorus she vigorously proclaims "I've got nothing to prove, I was once as miserable as you." The song mines her disgust with the system, and she nonchalantly puts down people in her new hometown, Los Angeles, who tell her they're in "the industry," by asking "Oh, are you in steel?" Her efforts to pitch her new album (which included this song) didn't get very far.

Sobule was bursting with ideas for songs that reflected her recent move from New York to California, and found a group of musicians who were eager to play with her. She didn't want to follow the desires and whims of a new record company, but without the hefty financial backing a label could offer, she faced a daunting challenge—how to raise the money to record and market a new record that she would produce herself?

Borrowing a trick from the public radio songbook, Sobule decided to ask her fans for donations online. "I had the idea a couple years ago," Sobule said just after she released her album *California Years* in the spring of 2009. "It's not a huge, huge fan base, but it's a mighty fan base. They're very loyal, and I think part of it is because I always write back." Sobule is the kind of performer who is happy to listen to her fans. "When they come to see me, and someone shouts out a song, I'm accommodating. If I like the song and I remember it, I'll do it. They just know there's a good relationship. So a couple years ago I e-mailed and said 'What do you guys think about becoming my record company?' And everyone seemed positive at the time. I finally had enough nerve to create Jillsnextrecord.com and actually really do it."

Figuring out how to structure the deal was the next hurdle. It was way too complicated to set up a system where fans could invest. But donations would work. "First, I had to think about how I would do it because I didn't want to say 'Well, just give me money.' It had to be in return for some gifts or services. It was a very public-radio idea I had. So I had a couple glasses of wine and went over the different kind of levels of donations and what they would get."

Donations started at $10—the Unpolished Rock (but with potential) Level, which allowed a donor to download the album for free when it was released. For the $200 Bronze level, a donor got free admission to all Sobule's shows in 2008; at $500, donors got a mention on the recording: "The last song of the record is called Donor's Song where your name is in a song." Double that and Sobule "would write you a theme song. And I had house concerts." On her website, Sobule recommended a $5000 gift, writing "I've played many house concerts where the host has charged his guests and made his money back. I'd go for this if I were you."

"The best one was—and this was a total joke, I didn't think it would happen—Weapons Grade Plutonium, where you get to sing on my record." A software developer from England surprised Sobule by taking her up on the offer, and she sings "La La La La La La" as backup on Sobule's song *Mexican Pharmacy.* "She did a bang-up job, I must say."

The website launched in January of 2008, and in less than three months, Sobule raised $88,969 from 638 fans, who posted enthusiastic notes in the website's guest book, including Mike G. who wrote, "For years I've been handing out business cards that say 'producer' on them. Of course it was a bold-faced lie . . . Until now." Sobule was thrilled, and a little shocked, at her success. "I would've done it no matter what I got. But it could've been humiliating; I think the surprising part was how great everyone was. And how they continue to support me now."

The backing of her new "producers" gave Sobule the cash to hire superstar Don Was (who won the first of many Grammys for producing Bonnie Raitt's hit *Nick of Time* in 1989) and a backup band of great musicians. But the donations didn't just go to recording the music. "That's probably half of it. It was really to try to do what a record company does—marketing, publicity, distribution. So even when I had my fantasy people on the record, they definitely gave me the poor artist rate."

How did the experience compare to making an album for a record company? "For the most part it was better," Sobule said.

"The scary part was—you better make sure the people who donated liked it. It was way more intimidating than working for The Man. I'd rather have them drop me than the people who actually buy something or download it for free and go to a show." Sobule released the album just a year after setting up the donor experiment. The novel funding arrangement certainly helped with publicity for the music—Jillsnextrecord.com was covered in stories from the Associated Press, CNN, *Entertainment Weekly*, and newspapers around the country.

California Years begins with the pretty "Palm Springs," following Sobule as she drives through Southern California in her Prius, on a quest to find the places "I've been searching for." The refrain is "Something's gonna happen to change my world." Sobule has made that something happen all by herself, taking all the aspects of making a record into her own hands.

MODERN ALCHEMY

In the spring of 2003, *Studio 360* took a trip up the Hudson to visit Dia:Beacon, a contemporary art museum that had just opened in a renovated 1929 factory building about an hour north of New York City. Kurt strolled through the cavernous galleries with Dia:Beacon's director, Michael Govan, who explained that the space, which had originally been a box printing factory for Nabisco, was perfect for a museum because the roof was filled with skylights; it had been designed so that the printers could see their work without the use of electric lights.

Later that summer, my husband and I visited the museum with our two sons, then five and eight years old. The boys were vaguely interested in the neon sculptures of Dan Flavin, and barely glanced at the obsessively beautiful paintings of Agnes Martin. But we needed to hold them back from clambering over the guardrail and into the huge circles and squares cut deep into the earth by Michael Heizer.

The familiarity of other materials intrigued them; they skirted carefully around Robert Smithson's *Map of Broken Glass*, played peek-a-boo through John Chamberlain's tall *Privet*, made of ribbons of brightly colored car parts, and wanted to snuggle up to Joseph Beuys's mounds of dark brown felt.

But one work in particular made them run and dance with delight: the giant *Torqued Ellipses* by Richard Serra, three twelve-foot-tall, mysterious, Cor-Ten steel sculptures tucked away in a downstairs room with huge windows. The boys wandered into the narrow entrance of each towering, rusted orange ellipse, investigating the interior space, hiding from us and giggling as we found them at the end of a long spiral, shouting to hear their voices echo on the curved walls. They loved these sculptures made out of industrial steel, and they instinctively knew how to play with them.

"Kids get it immediately," Richard Serra said about one of his *Torqued Ellipses* that was defining the sky in a beautiful corner of the garden at the Museum of Modern Art in New York City as part of his retrospective. "What's great about it is that kids will grow up with the notion that this is sculpture. They won't have to be told. Kids immediately get into it, and they play with it, and they immediately understand that it's in relation to their body and how they understand space, and how they map it."

Artists in *Studio 360* have often talked to us about the allure of transforming ordinary, everyday elements into powerful works of art. Here, three masters of this modern alchemy describe their passionate engagement with their raw materials. Ben Burtt, who designed the sound for all of the *Star Wars* movies, reveals the familiar sources he mines to create unforgettable sound effects for everything from Darth Vader's ominous breathing to WALL-E's wistful voice. Choreographer Elizabeth Streb pushes the boundaries of the human body's response to gravity, transforming her dancers into "movement technicians" who can fly, if only for a few breathtaking seconds. And Richard Serra describes how he shapes tons of steel into geometries never before imagined.

Burtt, Streb, and Serra articulate the intention behind their work, how the material inspires them, and how they strive to transcend the limits of what that material can do, whether it is noise or steel or flesh. These artists may not change lead into gold, but they combine, reshape, and lift their raw materials out of their familiar contexts, transforming sound, bodies, and metal into art that changes the way we perceive the world.

Sound designer Ben Burtt

Turning Noise into Sound

"The mismatch between movie sounds
and real life made me stop, and wonder,
and look into the idea that you could create . . .
a caricature of real sound for dramatic illusions."

Conjure up the image in your mind of young, blond Luke Skywalker, lightsaber in hand, dueling with the imposing black-masked Darth Vader. Now think about the sounds that accompany that vision. No doubt you hear the thrumming whoosh of their lightsabers as they attack and parry, something I heard imitated a million times as my children and their friends battled in the living room with their very own plastic blue and red lightsaber knockoffs.

The menacing hum and sizzle of the lightsabers were the very first sounds that Ben Burtt created for *Star Wars* back in 1977. When George Lucas asked him to be the sound designer for his movie, Burtt was still in graduate school at the University of Southern California. Once he saw the production paintings of a lightsaber battle, "I had in mind something almost immediately,"

Burtt told Kurt in *Studio 360*. "I was a projectionist at the time at the USC cinema department. In the projection booth there was an interlock motor on the projectors that made a wonderful hum. It was a musical hum, and sounded like a lightsaber to me. So I recorded that sound."

But the projector's hum alone didn't sound quite fierce enough to him, and he kept hunting. "I was doing some other recording in my apartment. I had a broken microphone cable which, when I carried the mike past the television set, picked up 'buzz' from the TV picture tube. Just the kind of thing you normally would not want in your recording. You'd reject it. But I thought, 'Oh, that buzz sounds dangerous.' So I combined the buzz with the hum of the projectors. The two together became the basic lightsaber sound."

Go back for a moment to your mental image of Darth Vader and Luke Skywalker battling—and you'll realize that Burtt still had work to do. A static hum, one that remains at a constant pitch and intensity, no matter how dangerous it sounds, can't convey the drama and motion of the lightsabers as Luke and Darth Vader wield them in combat. "To get the illusion of motion," Burtt said, "I played the sound over a speaker and then rerecorded it with another microphone. By waving the microphone back and forth in front of the speaker, you get a pitch change called the Doppler effect, creating the motion of the swords. And that's basically how it was born."

A few weeks after recording the lightsaber's hum, Burtt went on vacation with his family. "We were hiking under a guy wire at the top of a ridge in the Pocono Mountains in Pennsylvania, and a pack frame caught and plucked it. I said, 'Wow, that sounds like a laser gun.'" Burtt just happened to have his Nagra tape recorder with him on the hike and recorded the zing of the wire. "I struck it with pieces of metal, a wrench, a wedding ring, and various other things. Those recordings became the basic sound for all the blasters in *Star Wars*."

"Many of the most useful sounds I've used were discoveries made by accident. I would be going about some other business but then hear something interesting. I learned to keep my recorder nearby. I'd grab any sound that caught my attention. Even if you don't know what it's going to be. If something interests me, if something is provocative in some way, if it has a sense of suspense to it, or power, or humor, then it's worth gathering and stockpiling until I find a place in a movie to use it. I have a big collection and it's always growing. I'm probably always listening to what goes on around me, whether I want to or not."

From a very young age, Burtt was extraordinarily attuned to everyday sounds. "As a child, I remember playing with my grandfather's shortwave ham radio set in the blisteringly hot attic of his Ohio home during summer visits. I loved tuning between stations and listening to all the tones and beeps and whistles and static. Often I will do that today. I'll turn on a shortwave radio, put it next to the bed, and mistune it somehow so I'm really not hearing any station directly. There's something cosmic and enchanting about the endlessly different textures of random noises and tones—I find that it opens my mind. I find peace and excitement at the same time. So I guess that's something that I'm always going back to."

Burtt also had an early passion for the movies. When he and his family went to drive-ins, Burtt would put his tape recorder next to the speaker hung on the car window so he could record the film soundtrack and play it back later. "Of course I love the things I grew up with—*Forbidden Planet*, *War of the Worlds*, *Flash Gordon*, and, later, *2001*. That was a film with minimalistic sound effects, but very powerful dramatically. My favorite sound effect is the arrow from the *Adventures of Robin Hood*, 1938. It's something that just struck me as a kid. It was one of the reasons I got interested in how movie sound was different from the real world. If I went outside and fired my own little toy bow and arrow, it wouldn't sound anything like the one in the movie. It was the mismatch between movie and real life that made me stop, and wonder, and look into the idea that you could create sounds, perhaps magnify their qualities, and give them enhanced character for use in a specific movie. I discovered that designing movie sound is not just reproducing literal or natural sound, but most often creating a caricature of real sound for dramatic illusions."

Burtt's own iconic creations have no doubt sparked a similar curiosity in younger moviegoers. Think of the crack of Indiana Jones's long whip—which Burtt crafted from the sound of a Harrier jet fighter flying overhead at the annual Oshkosh air show. Or the ominous, labored respirations of *Star Wars* villain Darth Vader, which began with Burtt recording his own breathing through scuba gear.

When he was growing up, however, even though he made his first movie at the age of ten, Burtt never thought of movies as a career; he wanted to become an astronaut and go into space for real. At Allegheny College in Pennsylvania, he majored in physics and continued playing with sound. One summer, he made an adventure movie about World War I aviators, which won a national student contest, and Burtt was offered a scholarship to the film school at USC. "The hobby of filmmaking and sound recording kept looming up and eventually captured me. I went into space in the imaginary realm instead of the real thing."

Yet even Ben Burtt can get tired of science fiction. "When I finished *Revenge of the Sith*, after twenty-nine years and ten months of *Star Wars*, one of the things I was relieved about was that I would not have to create any more alien voices or robots, because they're so incredibly hard to do." A short time later, Burtt was invited to discuss a new animated film in development. "I listened to a pitch by Andrew Stanton, the writer and director of *WALL-E*. I hesitated just a little and then I thought, 'Oh, this is such a charming, fun movie. How can I say no?'"

WALL-E (the acronym for Waste Allocator Load Lifter—Earth Class) is a forlorn robotic trash compactor, left alone on an abandoned Earth covered in mile-high heaps of garbage. One of Burtt's major challenges in this movie was to create a compelling vocal character that would convey the curiosity and personality of WALL-E, a robot that, when not collecting trash, watches an ancient videotape of a made-up movie musical, over and over, and dreams about love. "I feel a lot of stress when I create new voices, I know that audiences hear voices differently than they do sound effects; they're much more critical. We're all experts at the nuances of speech and we can analyze that in a special way. The audience is much more critical of voices and thus the illusion is harder to create. It's easier to do a big gun battle and create lasers and spaceships. When I looked at the job I had for *WALL-E*, it was a bit daunting. I had to come up with convincing expressive sounds for voices, and the characters were going to be in every scene in the movie. The voices have to be consistent, and listenable too, something you like to hear."

As soon as boxy, rusty WALL-E rolls into view in the desolate landscape and lets out a robotic question mark, a moviegoer may immediately think the little robot is a distant cousin of *Star Wars*'s R2-D2. The two robots look nothing alike, but they do share some genetic material; what gives resonance to both characters, underneath the whistles and mechanical squeaks and whirs, is Burtt's own voice. "The successful robot voices have been those that always had in them a disguised human voice, giving the character what I call a 'soul.' It's one thing to create a machine voice out of sound effects, but often it is cold. On the other hand, you don't want to use a straightforward human voice; people recognize it as such. The trick is to blend the two, to have a human component present so that the character has some soul, but disguise it and process it in such a way that it has an electronic or machine-like character. Coupled with that is the addition of lots of sound effects, the motors or the tonalities of the character."

After director Andrew Stanton described his ideas for *WALL-E*, Burtt tried to translate that vision into a set of sounds. He spent about six months coming up with possibilities, sending them to Stanton, and then modifying them according to the director's feedback. "I even got to the point where I would sometimes put together a robot voice and the voice would say something like 'Hi, Mr. Stanton, I'd like a part in your movie.'" All of this exploratory work went on while the film was in early development, before there was any animation. "In conventional animation, if there are voices, they're recorded first and edited together, and then the animators have a guide to the exact timing and performance of the characters. We took somewhat the same approach here. I would create little sets of sounds, give that to the animators, who would invent animation to go with it. I am amazed at what they can do with just the animation; they can bring out so much of a character just through a pose, the angle of a look from one character toward another. We worked back and forth, doing these vignettes: I'd create sound; they'd create animation. I'd then modify the sound based on what was inspired by the animation, and vice versa, until everyone understood the sonic nature of each character, and then the story and the real animation could proceed."

Voices were just the beginning of the challenge; as he did for *Star Wars*, Burtt had to create the entire sonic landscape for *WALL-E*, coming up with sounds for everything from the spaceship's pneumatic doors, to lethal laser blasts, to the atmosphere of an abandoned Earth. To portray a gust of wind echoing down a giant canyon of towers built from trash, Burtt dragged a punching bag across a carpet and recorded that sound. When WALL-E drops a compacted block of garbage and it bangs on the ground, "that's actually a pluck on a harpsichord that I recorded one day when I picked up my daughter from her music lesson."

Burtt scrounged around his own bathroom to create the buzz of a twitchy, hyperactive scrubber robot named MO. "His prominent feature is a little cleaning brush, which is my electric shaver from home. His little voice is a sped-up version of me doing sounds I might have made in a third-grade class and got punished for. But now I get paid for it."

When he began crafting sound for the movies in 1977, Burtt could have chosen to manufacture them in the studio, the way sound had been developed for cartoons. Later, using computers, he could have generated just about any sound he wanted. But Ben Burtt is clear that "despite the digital age, I still emphasize field recording and organic sound from real physical objects. I think that the audience, even if they can't identify a sound, they associate with those sounds some kind of reality, so when you use those acoustic sounds in a science fiction or fantasy film it really helps to sell the fantastic as being real, because the sounds that are there sound comfortably real."

Choreographer Elizabeth Streb

Changing Bodies into Projectiles

"It's necessary to put yourself in harm's way . . .
out of your comfort zone . . .
[to] discover new physical territory."

Some sounds carry powerful emotional cargo, like the sound of breaking glass. You have likely, in your life, heard it more than once—perhaps you've dropped a juice cup on the kitchen tiles, or your neighbor has hit a baseball through your window. Imagine the sound at the moment the glass shatters. Now, imagine it's not a baseball that has smashed through the window—it's a human being, diving headfirst through a pane of glass suspended three feet above the floor.

You've just visualized one of the signature moves of the STREB company, a performance troupe founded by choreographer and self-described "action architect" Elizabeth Streb. She has devoted her career to challenging the boundaries of what humans are capable of, using bodies, glass, and metal as her materials. "I wanted to see the effect of action on substance, so we have this pane of glass and we actually dive through [it]," Streb explained. She said that this particular dance is "very short, it just takes the time it takes to dive through the glass, but usually the person who's diving gets a little cut, usually the glass flies into the audience."

Streb first performed this move at a benefit for the Joyce Theater in New York. "Standing there, looking at the glass in this theater, I thought, 'Oh my gosh, I've got to dive through that glass.'" How had she rehearsed for that moment? She hadn't. "You don't really practice it; each piece of glass costs four hundred dollars, that's one reason you don't. Two, it doesn't mean just because you do it once you can do it twice. A lot of it is, do you have the heart, like a boxer, to go on? It was a phenomenal experience."

That kind of experience is what Streb and her dancers are constantly searching for, for themselves and their audiences. They rehearse in a former mustard warehouse in Williamsburg, Brooklyn, where metal trusses reach almost all the way up to the thirty-foot ceiling, a trapeze takes up one side of the fifty-by-one hundred-foot space, and gym pads, some four feet thick, cover much of the floor. Most days, the room is filled with the sounds of impact: bodies falling hard onto pads, or crashing against Plexiglas, or running into each other at full speed; high school students panting as they learn how to fly on the trapeze or do a hands-free cartwheel over classmates who crouch on a mat; little kids giggling as they jump on a giant trampoline during a birthday party.

This is SLAM—the STREB Lab for Action Mechanics. Part school, part rehearsal space, part circus, part gym, part theater, SLAM is a gathering place for passersby, students, and the "movement technicians" who perform as part of STREB. Elizabeth Streb conceived of this space as a kind of twenty-first-century urban garage, where all kinds of people can meet and create and watch her company rehearse and perform. It's home to her team of gymnasts, modern dancers, and trapeze artists, where together they develop the action events that have captivated audiences all over the world.

In college, Streb studied modern dance and was fascinated by the work of the choreographer Merce Cunningham. But her major inspirations go even further back—to the circuses she was taken to as a kid growing up in rural New York, and the televised stunts of the daredevil Evel Knievel. "He was living my dream and his own, too. He would do things like going over twenty buses. And then he went and did it in London, their buses were bigger, and he knew he was going to crash. He knew that. But there were five thousand people there, and he did it anyway." Streb loved riding motorcycles, and downhill skiing, and her passion for speed and daring is evident in her choreography.

She terms what her performers do "pop action," which is different from the kind of movement she learned in school. In pop action, a dancer moves from one position to another, from upright to prone, squatting to soaring, without showing any of the transitional steps. "I think when you're on your feet, right side up, transferring weight, that's what I call normal dance, you're not dealing with the forces of action. For us, we're dealing with rough-and-tumble gravity, like falling or slamming or dealing with impact or dealing with rebound, and so once you get into that training, I think that the issue of extreme physicality gets described in a different way than a more normative dance."

In a piece that beautifully illustrates the impacts and forces she describes, Streb pits humans against Plexiglas. An eight-by-eight-foot Plexiglas wall is secured flush with the front edge of the stage, and the dancers hurl themselves against it as hard as they possibly can. "We do a power run to up the ante for the speed, so when we actually get stopped by the Plexi, it's a huge explosion." In addition to body smashing, the dancers literally run up the side of the Plexi wall and push off into backward somersaults and short flights to the floor. The dance is just three-and-a-half minutes long, but the performers say it's the hardest thing they do; it's like running a three-and-a-half-minute mile. Everything—dancers, Plexiglas, the rest of the stage—is miked so that the sound of each impact is amplified and impossible to ignore.

The company employs various other materials to extend the possibilities for movement, such as the hamster wheel for humans, a metal contraption twenty feet high. Dancers run inside and outside the giant circle as it revolves, and fall straight from the top, body parallel to the floor, onto mats that don't quite look thick enough. A dancer once remarked that falling isn't the part he worries about: they've developed a technique to land with their ribs pulled in, so that they don't break. It's getting onto the wheel that can be scary, because as the wheel passes its triangular support, there's a potential for some "guillotine action." The performers make beautiful use of all parts of the circle, running on top, leaping over their colleagues who are grasping the outside rim with arms and legs and traveling 360 degrees, or standing inside the circle with heads toward the center as the wheel revolves. It's exhilarating to watch. "When you're a visual artist or someone making up lifetime events, your job is to figure out—'Okay, what do people want to witness?'" Streb said. "And it usually [means] that they're not going to be doing it, too. I think that the more extreme the moment is that anyone constructs, the deeper the watcher gets into it and participates in it."

This work takes a lot of physical courage and has inherent in it a big dose of risk. "We ask the body to do things that aren't really, in the general world's view, very healthy to do. If you're not in danger, or, I guess, willing to put yourself in danger, then what you're able to discover physically will be zero. So my job as a choreographer is to devise situations that at first I think are absolutely crazy, dangerous—like 'you can't do that.'"

There is one big rule—no landing on the top of your head. "Anything else is okay, you can land on any other surface of your body except for your head." The ensemble also has an unusual approach to the bumps, bruises, and sometimes broken bones that are all in a day's work: "Pain is something we use judiciously as a word because it backs you away from doing anything that's uncomfortable initially. We call it, actually, 'another interesting foreign sensation.' If you and I walked in there today and did [what the dancers do] it would probably hurt. But the more you do it the more you understand it has a technique, and it hurts less."

The strong, flexible performers come to the company from a myriad of backgrounds: modern dance, gymnastics, extreme sports, the circus. All share an appetite for intense physical experience. "What the audience sees is somebody taking their physical entity and just stretching it beyond their comfort," Streb said, but she emphasized that the allure of danger isn't the only reason for these feats. It's not that the dancers *want* to put themselves in harm's way, she explained. "It's the other way around. To effect certain actions onstage, it's necessary to put yourself in harm's way. I think it's a means to an end. If you don't allow yourself to get out of your comfort zone, then you can't discover new physical territory. And it comes with that territory that there's going to be certain things that scare you. You're going to try to locate them, and unravel them, and go further and further and further toward something that you couldn't imagine doing physically prior to that initial investigation."

When they are onstage the dancers may look fearless—but that's part of the act. "We are all normal people and we're scared of similar things that other people are scared of. We just have an appetite to dig into them and conquer them on some level. And then the fear switches. Let's say you're twenty feet up and you're gonna tip off and just fall, and your job is to stay perfectly horizontal, before the ground comes. And some days it's no big deal to do that. All of a sudden, on a Wednesday, you're just terrified, you can't do it. There's no accounting for those shifts in fear."

Streb has had moments when she's questioned how far she's willing to go. Okay, she will admit she has had one moment, when one of her performers, Terry Dean Bartlett, wanted to fall farther than they'd ever tried before. "We were thinking, 'If you can fall flat on your face from eight feet, you could probably do it from ten. I bet you could probably do it from fifteen. Why not just go up to twenty-five?' It was the first time I ever said, 'No, not on my time, not in the show.'" Bartlett persevered, and ended up in performance falling from twenty-five feet in the air, belly-flop style, flat on his face. Each time, Streb said, the entire audience gasps at exactly the same moment, when he's about halfway down. "It's one move, and it is remarkable. So now we're thinking, 'Thirty feet? Maybe fifty?'"

Teaching people how to fly, even if it's just for a few gasp-inducing moments, has been a motivating force for Streb. "I'm very interested in the concept of flying, and a lot of this physical technique over the last twenty-five years has come from that." She remembered when she decided to see if she could fly. "One time, twenty years ago, I said, 'Well, just climb a ladder and hurl yourself into space.' And then I crashed into the ground, because I hadn't planned ahead enough to realize, 'Oh, I'm going to have to take the hit.' I just thought, 'I'll see what happens.'" Since then, she's developed techniques to minimize the damage, including thick gym pads and bungee cords, but she's continued to play with the idea. Some pieces are simple and brief: a long cord connects two dancers to each other, and one moves suddenly, jerking the other off her feet and into the air for a split second. Other dancers gracefully swing around a tall central pole, their bodies suspended in harnesses, or the performers tether themselves to an apparatus that lets them dance up a wall. Streb loves machines almost as powerfully as she loves movement.

With this complicated toolbox of mechanical contraptions and strong bodies, Streb has been able to pursue her dream of destroying what she calls "the tyranny of the floor." When you experience a STREB performance, your perspective is suddenly shifted; watching the dancers run up the Plexiglas wall, you feel you are under the floor, looking up at their pounding feet. As dancers do cartwheels on the wall, suddenly you imagine you're suspended in midair above them. When they bound into the air from a large trampoline, for a few breathless moments you can imagine flying with them as they fall to earth, arms outstretched like birds.

The MacArthur Foundation awarded Elizabeth Streb one of their so-called genius grants in 1997, commending her "gravity-defying movement." Streb loves to play with gravity, but at heart she is a realist—she has said she doesn't want to defy gravity, because if she did, she would definitely lose. Instead, her work is more like a romance with that elemental force. "Gravity exists. And the camouflage of gravity has never been okay for me." Getting humans to fly is her goal, but so is the crash when they hit the ground. "I want the beauty and the beast, I want them both." And, perhaps paramount, Streb wants to create an experience that will thrill both her dancers and her audience. "Our job is to not die in front of them, but their job is to want to see how far you're willing to go. Otherwise they can sit home, or walk down the street, or sit on their stoop and see life go by. I want to make a virtuosic moment happen that they're not completely comfortable with."

Sculptor Richard Serra

Transforming Our Experience of Space

"I really wanted to involve myself with ancient materials."

The forces of gravity are an integral part of Richard Serra's monumental steel sculptures, each of which weighs tens of thousands of pounds. Yet he never bolts anything to the ground; every huge section of every sculpture is entirely freestanding, designed to balance on its slim edge.

In the spring of 2007, the Museum of Modern Art mounted a retrospective of Serra's work and created two videos to document the careful and painstaking process of installing his sculptures. In just a minute and a half, one video captures the two days it took to place his *Torqued Ellipse* and *Intersection* on the white marble terrace in the garden. The video is full of furious action as a large crane lifts giant sections of steel over the high garden wall, figures of people scurry around to place them carefully, the sun casts quickly moving shadows, and cars race up Fifty-second Street as night falls over the garden. The time-lapse video beautifully illuminates the delicacy in this monumental performance of people, machines, and steel.

The second video begins with a gigantic garage door opening into a cavernous gallery that was designed specifically to support the tremendous weight of Serra's sculptures, as the artist describes the process of erecting the work he created for this large space. Serra has a deeply lined face, framed by wild gray eyebrows and closely cropped white hair. In his green work jacket with gray hooded sweatshirt underneath, he easily blends in with the cadre of workers who use forklifts and a giant gantry to carefully guide the huge curved slabs of steel onto their marks on the floor. "Every piece is balanced," Serra describes in the video. "The problem is when you lift them—if you don't lift them from the right pick points, then they're out of balance. That's when you have to really pay attention to get them back in balance before you set them. Once they're set they're weightless, but while you're moving them they're not. You have to watch everything all the time. You're lifting a lot of weight onto a platform: there are cranes; there are sleds; there are jacks; there's a lot to pay attention to. It's hundreds of tons coming in here."

Serra's relationship with steel began during his childhood in San Francisco, when he lied about his age (he was just fifteen) to work for United States Steel. He spent all his summers through college in the steel mill. Yet the sculptor said that "if someone had told me when I was catching rivets that I would go on to become a sculptor and be interested in steel, I would have said, 'I don't think so.'"

It's not that Serra didn't think about being an artist—from the time he was a little boy he knew that was his destiny. Or, perhaps better said, his mother knew it was his destiny. "I don't know if you know anything about Jewish mothers, but they're very important. And she was very insistent right away. I think in the third grade a teacher pulled her into class and they had all of my drawings up around the room. She called my mother in and pointed it out and said, 'You should take this child to museums to encourage what he's already doing.' So my mother totally got onto the program and started taking me to museums very early and introducing me as 'Richard the Artist.' She told my older brother that he was going to become an attorney, and he's become a very famous attorney."

Sculpture was not at the top of his list when Serra went to college, however. He studied English literature as an undergraduate, and attended graduate school for painting at Yale. After school, he lived in Paris for a year, and then received a grant to paint for another year in Florence. While in Italy, Serra took a trip to Spain, where he visited the Prado and saw *Las Meninas*, which Diego Velázquez painted in 1656. The experience changed Serra's entire direction.

In the center of *Las Meninas* is the Infanta Margarita, a small blond girl wearing a white dress, tended by her ladies-in-waiting. They are standing in a large, high-ceilinged room with paintings covering the walls. Directly behind her is a mirror, and reflected in it are the faces of her mother and father, the king and queen of Spain. In the right corner there's a doorway through which you can glimpse a man in the hallway beyond. And to Margarita's right, standing next to a giant canvas, of which we can see only the back and its wooden stretchers, is the painter himself, palette in one hand, brush in the other.

Richard Serra said that as he stood in the Prado, looking at *Las Meninas*, "I realized that Velázquez was looking at *me*. In most paintings, the subject was depicted in the frame, but here was a reversal, where the perspective of the painting was outside of the painting, as if a wall had fallen down and there was a transparent box. There was a projection forward, where I was part of the painting. And I thought, 'I am the subject.'

"I thought I couldn't do that in painting. I was at the point where I was using a stopwatch and painting squares out of randomness, and wasn't getting anywhere. So I went back and dumped all my paintings in the Arno and thought, 'I'm just going to start playing around.'"

After destroying his paintings and giving away his brushes, Serra "started stuffing animals, started living with animals, started doing surrogate zoos, without the faintest idea of what I was doing." For his first solo exhibition, in Rome in 1966, Serra created boxes in which he kept live animals. "I wanted to get rid of my education; I wanted to do something other than make art, [something] that was completely different from how I had been taught. And I wanted to forget about the conventions of art."

He moved back to the States, found a studio in Manhattan, and abandoned his "barnyard surrealism," keeping his eye out for anything that might spark his imagination. One day he heard about a warehouse filled with rubber that was going to be thrown out. "I phoned the CEO and I asked if I could take it. He said, 'As much as you can haul away, you can have.' So I got Chuck Close and Phil Glass and Steve Reich and Michael Snow—we had a small furniture moving company—and we moved all this rubber up to my loft. To me it was like being given a grant of material."

Once the rubber was in his studio, Serra had to figure out what to do with it. He turned to his experience at Yale, where he had studied with the influential modernist painter Josef Albers, who "taught you procedures in relation to matter. And one big subtext of the lessons was that matter in itself informs form. So if you make something in rubber, if you make something in lead, you make something in cement, you make something in steel, you can make it exactly the same way but it has a different readout, a different reference, a different psychological hit." Serra thought about the architect Louis Kahn, who famously said, "When you are designing in brick, you must ask brick what it wants or what it can do. Brick will say, 'I like an arch.' You say 'But arches are difficult to make, they cost more money. I think you could use concrete across your opening equally as well.' But the brick says, 'I know you're right, but if you ask me I like an arch.'"

"If you go back to someone like Kahn—'I go to the brick and ask what it wants to be'—all of that was lost in the sixties, when [art] became much more theoretical and was divorced from matter," Serra said. "And for me, that was the path I didn't want to take. I really wanted to involve myself with ancient materials. So what I did was I wrote down a verb list, there must have been over a hundred of them, and I started to enact the verbs in relation to matter. What it did was free me from the idea of 'Oh, I have to make art.' I'm not going to make art, I'm just going to involve myself with process and procedure in relation to matter or material and see what will come of it."

You can feel the English lit major come through in Serra's list of verbs, which reads like a poem of directions for what he could do to his materials, beginning with "to roll, to curve, to crease," and ending with "to continue." By instruction number four, "to lift," Serra realized he had hit upon something important. "I took a piece of rubber that was about, oh, four feet wide, twelve feet long, and took it on its . . . center edge, and lifted it up," Serra said. "And what happened is, it formed a continuous topological form. I thought, 'Oh, jeez, it's freestanding, this is interesting, and it has a volume like sculpture.' You do these things and you think, 'Oh, lay it back down and try something else.' And I had that one up long enough to come back day after day to think, 'Well, I could put my name on this one, right?'"

To Lift, from 1967, is in the collection of the Museum of Modern Art in Manhattan. It looks like a molten black teepee, or a dark serape just vacated and standing up on its own. Simple, curvy, and eloquent. And powerful enough to get Serra to re-evaluate what he was doing. "I thought that up to that point I was making art. It's very easy to call yourself 'an artist,' and in a general sense I think that's what every young person wants to do. To say that you're going to be a 'sculptor' means that you're going to limit the field, you're going to enter the academy of a convention. You think, 'Whoops, am I really going to do that, am I really going to take this one on?'"

While living in Paris, Serra had spent time in Brancusi's studio. "There was something interesting in the way he cut into space with a line. When you're young, Brancusi is a kind of a handmaiden to moving any way you want." Serra was a little in awe of Giacometti, whom he would see when Serra and his friend Philip Glass hung out at La Coupole in Paris. Giacometti "used to come in at one in the morning with plaster in his hair." Serra understood the tradition he would be engaging if he became a sculptor. "And I thought, 'Well, look, I can't get around this. This is freestanding, you can walk around it, you can look into it, you can look through it.' I put it up. I can't say, 'It's not sculpture, I'm just an artist.'"

He made the journey from artist to sculptor "through defining procedures for myself in relation to material and what I could do with it, in relation to physical process." At the same time, he noted, he was freeing himself from the constraints of making art, by approaching it as play. "I'm interested in the notion of play; [I'm] not interested in the end, [I'm] interested in the activity itself; [I'm] interested in not worrying in a self-conscious way about what I have to make."

Industrial materials were his new playthings. Over the next several years, Serra hung ribbons of rubber and neon off hooks. He ladled molten lead on walls and floors, and made a short film, *Hand Catching Lead*, which is exactly that: the artist's outstretched right arm and hand grasping to catch small rectangles of lead as they are dropped from above the frame, his fingertips getting dirtier and dirtier each time the lead is caught. In his studio, Serra propped huge lead sheets onto walls and held the sheets up with long, heavy lead pipes.

In 1969, he again moved away from the wall to create *One Ton Prop*. "It's four plates, about two hundred and fifty pounds, three hundred pounds each, tilted like a house of cards, so that the corners overlap as they go around. And the compression holds them tectonically in balance. But there's a tendency for it to implode because it's lead, and lead has a low order of entropy, so finally the piece will collapse if you don't pay attention." Serra gave the sculpture the nickname *House of Cards*.

In a photograph of *One Ton Prop*, which looks like it was taken in Serra's studio, you can see where someone has written "Dick's Art" with a fingertip in the dusty patina of one of the lead sheets. Leaning together, they are imposing, even though they're just four feet tall, and a little unnerving, because you immediately sense the precariousness. Serra said he came to this arrangement through an exploration of the inherent capabilities of the material. "My intention was to get to the basic tectonics of building. What's the simplest way to build in terms of balance? What's the simplest way to build in terms of stasis? I wanted to get into tectonics and figure out how can you structure something with no joints? How can you structure something without fastening it? You put together something without a weld, how can you have something self-sufficient within its own logic of building? That's what it was about. It wasn't about, 'Oh, this is going to fall over and hurt somebody.' "

In his next work, *5:30* (named for the time of day when the sculpture was completed), he took the box and opened it up, with four sheets of lead making two corners, separated by a space, so that the interior was now visible, a lead pipe on top holding it all together by means of placement and weight. Not quite large enough to walk inside, and still too precarious to make that advisable. But you can see a connection between this sculpture and the next phase in Serra's exploration of industrial materials: steel.

Serra's mantra is "Work comes out of work." He said that once he makes one sculpture, "it leads you to a problem, and you work on a series of problems, and often the solution to one idea leads to a different idea." As he was creating ever-larger sculptures, he didn't "have some grandiose idea about using heavy tonnage of steel. In fact, going from lead to steel was another convention I had to swallow and decide I was going to take on."

Earlier in the twentieth century, other sculptors had embraced iron and steel. Julio González, a Catalan sculptor, had welded iron into dramatic sculptures that often evoked mythical characters. Pablo Picasso studied with González, and in turn created his own iconic works out of the material, including his *Chicago Sculpture*, a landmark in that city, which looks like a huge head, both human and simian, cut and welded out of Cor-Ten steel.

Serra knew that history, and rejected the methods of cutting, folding, and welding the metal. Instead, he looked to his experience catching rivets in steel mills and his father's work in shipyards. "I thought, 'If I'm going to use steel, I'm going to use it the way I was raised, the way I knew.' I moved back into the Industrial Revolution and took the nature of steel and thought, 'What are the properties of this material, and how can I use it in sculpture, in ways that it hadn't been used in terms of its potential for tectonics and its potential for building sculpture?' "

Serra also made happy use of accidents. In 1972, he moved a two-by-four-foot plate of steel that he was using as a backdrop for his lead castings out of the way, pushing it into a corner of his studio. With one edge shoved into the corner, the whole thing stood up, jutting into the room, and Serra thought, "Now, isn't that interesting!"

Soon after, a gallery invited Serra to make work for a square room, and he created *Circuit*, a sculpture that bisects the room with four towering plates of steel jutting into the center, from each corner. This giant "X" of steel, which doesn't quite meet in the middle, draws you into the room but also gives off a feeling of impending danger similar to *One Ton Prop*; the sheets of steel, twice as tall as the viewer, are not bolted down but held up entirely by their edges, one edge at the juncture of the corner, the other on the floor.

As he described in the MoMA video, when placed correctly, Serra's sculptures are completely stable. But over the years, there has been tragedy as works were erected or taken down. In Minneapolis, a two-ton sculpture collapsed while being installed in 1971, killing one of the workers. The company that fabricated the sculpture was eventually found responsible for the accident, but the public reaction against Serra was, as he put it, "vicious." He later said, "I was harassed, ridiculed, disgraced, and was told by friends, other artists, museum directors, critics, and dealers to stop working." Serra went into analysis, and began working in Europe and Asia—"anywhere where I could find support." He has acknowledged the risks involved in putting up his enormous work: "Rigging is a dangerous business. Danger is something that surrounds that activity. The sculpture when it's erected is not dangerous."

There was controversy, too. In the late 1970s, Serra was commissioned by the General Services Administration to create a sculpture for the Federal Building in lower Manhattan. *Tilted Arc* was installed in 1981. One hundred and twenty feet long, twelve feet high, the sculpture curved across a circular courtyard between Lafayette Street and the massive Federal Building, casting beautiful shadows on the cobblestones. Not everyone was enamored of the project; the high wall of tilted steel discomfited some viewers, and there was a vocal outcry, led by a judge who hated walking around the sculpture on his way to work. The critics eventually got the sculpture removed, over the objections of many in the city's cultural community. The battle was traumatic for Serra, as well as for New York's art world.

In the 1990s he began exploring tighter curves, enclosing space as well as redefining it in his series of *Torqued Ellipses*, the sculptures that so entranced my children at Dia:Beacon. They're the same height as *Tilted Arc*, twelve feet tall, but here the steel curves around itself to create an ellipse on the floor, and another ellipse at its upper edge, which is set at an angle to the one on the floor. Before Serra figured out how to bend steel into those curving shapes, no one had ever created such a form.

Initially, Serra thought that someone had—namely, the architect of the Church of San Carlo, which Serra had visited in Rome. "I looked at the floor, which is a beautiful ellipse. And I looked at the ceiling at the side aisle—a beautiful ellipse also. I misread it. I thought, 'Oh, the ceiling is on an angle to the floor, it's not this thing that rises perfectly in its elevation and one plane is on top of the other.'" But when Serra walked into the center of the church, he realized that he was wrong. The church was an ordinary elliptical cylinder that rose straight up to the ceiling. (Imagine an elliptical ship's funnel and you can picture the shape, just much, much larger.) The sculptor became fascinated by his misinterpretation of the space and wanted to try to make what he had imagined, an ellipse that twists as it rises.

So he called a nautical metal fabricator who was working with the architect Frank Gehry. "And I said, 'Look, what if I have a void on the ground—an elliptical void—and I have an elliptical void (let's take the most extreme case) at a right angle to it. Just air. I want to put a skin on it—paper, steel, whatever. Can I do that without changing the radius?' And he said, 'I've never seen it before, I don't think you can do it, but we can't play with you right now anyway because we have to build [Gehry's design for the Guggenheim Museum in] Bilbao, so it's not going to happen.'"

Serra was determined to make it himself. "So we took two pieces of wood shaped like two ellipses—two oval wheels—and we put them at a right angle to each other, connected by a rod." Imagine two bicycle wheels that aren't round but are elliptical, with the long sides at right angles to each other. "Then we took a piece of lead and we laid it out on the ground, and we took this wheel that wouldn't roll and we rolled it, from side to side because it wouldn't roll, wrapped the lead around it, cut top and bottom off, and we said, 'Oh, we got this thing!' Then we unfurled it after we wrapped lead around. When you unfurl it, that's the shape." Serra sent the lead model to the engineer who had said he couldn't work with him right now. "I said, 'Is this close to what we need to make?' And he said, 'Did you use a CAD2 [a computer design program]?' I said, 'No, we used two pieces of wood and a stick.' And he said, 'Oh, we can play with you tomorrow.'"

Serra has gone on to torque toruses (a torus, loosely speaking, is a doughnut shape) and to torque spirals; for his retrospective at MoMA he made *Sequence*, which is two torqued ellipses connected by an "S." From above, it's a beautiful double-"S" shape—but when you walk inside it, it's a strange and wonderful experience; you become a bit disoriented, not sure of your direction or whether you're retracing your steps, as the twelve-foot-high steel walls on both sides of you bend, either toward each other or away. "Basically, you walk from one ellipse to the other through a spiraling section, almost like a big serpentine curve that leans outward and inward and reverses itself as you walk," Serra said, "so there's a point there where you actually change directions. The piece is very much toward you in one direction and then it falls the other way. It's very fluid and fast. In terms of the tonnage—that's not the issue. Your body gets [involved] in the speed of the piece and your relationship to walking. You adjust to its moving. You're almost being propelled through. Because there's no vertical axis, you have to follow the weight of your body as it adjusts to the leaning of the curve."

When you are inside a Serra sculpture, you don't see the artist looking directly at you, the way Serra did when he first encountered Velázquez in *Las Meninas*. Yet because these immense steel works take time and motion to fully experience, Serra has succeeded in doing what he couldn't do in painting—he makes us the subject of his sculpture. "Once the sculpture got off the pedestal, and you became the subject, the subject and the content matter is your experience," Serra said. "This is your experience in relation to your walking. You don't have to know anything about anything. In fact, if you don't know anything about sculpture, it's probably better."

THE CULTIVATED AND THE WILD

For several years I produced a radio series presenting concerts from Carnegie Hall. I loved my job, but I think I may have loved my commute even more; most days I had the tremendous good fortune to walk to work through Central Park, enjoying the fantastic contrast between the skyscraper skyline and the huge oak and maple trees that line the park's rambling paths. On early-summer evenings, I'd choose a route home that wound past a small hollow near West Seventy-second Street where hundreds of fireflies would begin to blink just as the lights went on in the apartments on Central Park West.

More recently, through the windows of the commuter train that carried me to work each day at *Studio 360*, I found myself captivated by the rough and blurry borders between industry and nature. At first, the thirty-minute ride was a time to respond to e-mail or catch up on the sleep that is a precious commodity for a working parent of small children. But after a while I put down the electronics and simply stared out the window.

I especially love the path the train takes through the swampy marshes called the Meadowlands (probably best known as the home of the New York Giants's stadium, and where I have yet to see a meadow). The train tracks cut through cattails and at times are surrounded by open water on both sides. Under the huge cement supports of the New Jersey Turnpike, snarled train traffic into the rail tunnel under the Hudson often means long delays. But this allows more time to take in a vista of abandoned brick factories covered in brilliant graffiti, a pond punctuated by a flock of white swans swimming beside tall, tapering red radio towers, hills of landfill veiled in grasses, and lowlands covered with phragmites. That invasive marsh reed covers the landscape like the thick pelt of an animal, brown and feathery in winter, thick and verdant in spring—except for the gravel pits, truck graveyards, and refrigerators dumped by the side of the water. There's beauty on bright mornings when turtles sun themselves on abandoned tires half-submerged in the brackish swamp, while statuesque silver electrical towers march toward the rusted filigree of the Pulaski Skyway. This is a landscape where the industrial and the wild collide.

The friction between the natural world and what we have wrought within it fascinates many contemporary artists. For the poet Stanley Kunitz, the bounty of the garden he created out of a sand dune became his deepest source of inspiration. Conceptual artist Mel Chin believed he could sculpt the earth itself with plants; in doing so, he proved a new scientific theory about the use of vegetation to clean up toxic waste sites. Building on Chin's work, the landscape architect Julie Bargmann transforms old mines and factories into parks that offer natural beauty yet retain their industrial history. All three recognize that we are "partners in this land / co-signers of a covenant," as Kunitz describes it, and each transforms this responsibility into a canvas for exciting and unusual works of art.

Poet Stanley Kunitz

Pruning to Reveal the Essence of Trees and Poems

"At my touch the wild / braid of creation / trembles."

For more than forty years, Stanley Kunitz spent his summers in Provincetown, Massachusetts, in a wooden house on Commercial Street overlooking the bay. To reach the front door, a visitor had to negotiate a narrow path of crushed oyster shells that wove through a lush garden of ferns, flowers, shrubs, and trees. The poet planned the path carefully, making sure there were no straight lines, concealing the view so that the beauty of the garden revealed itself slowly. Kunitz approached his art the same way. "One of my principles is never to try to explain what a poem is about," he said in his final book, *The Wild Braid*. "That's a straight line to me. The path to the understanding of the poem is for me always circuitous, it's a winding path, and I think of the garden as a winding garden."

In the summer of 2001, when *Studio 360* was not yet a year on the air, Stanley Kunitz showed us around his glorious winding garden. He was ninety-five years old, the Poet Laureate of the

United States (for the second time—he had been Consultant in Poetry to the Library of Congress in the 1970s), and still tended the garden himself. "You're looking at rudbeckia over there," he said, pointing to brilliant yellow flowers as he walked through the terraced, verdant space that sat just outside the simple workroom where he wrote his poems, often laboring late into the night. "This one is called Diana; it has beautiful white blossoms. In front of that are varicolored phloxes. Plenty of bees; in fact, one has to be careful going through this garden because the bees have taken it as their own property and feel they have first rights. This is lavender; that does very well here, but of course the soil now is very rich. Anything will grow."

That wasn't always true. When Kunitz and his wife, the painter Elise Asher, bought the house in 1962, their front yard was a hill of sand, fifty feet wide and thirty feet long, sloping fifteen feet downward from house to street, with not a blade of grass growing on it. For two summers, Kunitz worked with a mason to create five brick terraces that held the earth back from washing into the street with every rainstorm. He brought in compost and collected seaweed on the tidal flats of Provincetown Harbor to enrich the soil, and then began the process of planting. In his last book Kunitz said that he "conceived of the garden as a poem in stanzas. Each terrace contributes to the garden as a whole in the same way each stanza in a poem has a life of its own, and yet is part of a progressive whole as well."

To anchor the garden he planted Alberta spruces in each of the four corners, which grew tall and dense, offering shelter to generations of garden snakes. "They love living inside, protected and free from intrusion," Kunitz told us. "In September, as the nights get very chill, they hang out from the branches, often intertwined, two of them together. At that stage they are very vulnerable to approach. What I love to do is stroke them as they're hanging there. And they learn through the years—I guess it's passed down through the family lore—that this man is not going to hurt them. They respond by quivering. That's one of the joys of living in this garden, certainly. I have a poem that I call 'The Snakes of September,' which deals with that and what it signifies to me."

It's wonderful to hear Kunitz read the poem in his gravelly tenor. He describes approaching the tree to find two snakes

> *. . . dangling head-down, entwined*
> *in a brazen love-knot.*
> *I put out my hand and stroke*
> *the fine, dry grit of their skins.*
> *After all,*
> *we are partners in this land,*
> *co-signers of a covenant.*
> *At my touch the wild*
> *braid of creation*
> *trembles.*

A couple of years later, Kunitz took that beautiful phrase, "the wild braid," for the title of his final book. The slim volume features poems from throughout his career, as well as entries that read like a diary, drawn from two years of conversations between the poet and his assistant, Genine Lentine, as they talked about nature and art while working in Kunitz's Cape Cod garden and his apartment in New York City. In one entry, Kunitz describes the juniper tree he planted near his front door, which weathered hurricanes but lost so many limbs that he decided to prune it as if it were a Bonsai. "As with the making of a poem," he said, "so much of the effort is to get rid of all the excess, and at the same time be certain you are not ridding the poem of its essence."

Photographs of the poet in his garden illustrate the book.

Kunitz was then nearing one hundred years old, and we see him stooped and wrinkled in a plaid flannel shirt and worn corduroy pants, lugging buckets, along with his cane, through long fronds of ferns that reach out to him from the border of the garden. In another, he is wearing a tan jacket and blue hat, sitting on a chair with his back to the camera, framed by towering red phlox. My favorite photograph is a close-up of the poet's hands, clasped behind his back, his fingertips dark with the rich loam of the garden.

Over his long life, Kunitz worked in many gardens. As a young man during the Great Depression, he bought a farm with an ancient, unheated farmhouse on Wormwood Hill in Mansfield Center, Connecticut, for $500 down. He spent three years fixing it up, getting to know all the wildlife on his hundred acres, including a family of owls that he patiently befriended, spending hours standing still underneath their nest until they accepted him as part of their world. In time, like the snakes in his poem, they allowed him to stroke them, and finally he carried the mother and four babies to his attic, where he had set up a branch for them. There they lived for a few years, flying in and out of an open window. "My encounter with this family of owls," he wrote, "was one of the most intimate of all my experiences with the animal world, a world I consider to be part of our own world, too."

Many of the poems in *The Wild Braid* evoke the world of plants and beasts. We encounter the imperious raccoons who claim their tithe of vegetables from his garden, and the crickets whose song makes the poet marvel "to hear so clear / and brave a music pour from such a small machine. / What makes the engine go? / Desire, desire, desire. . . ." Rather than anthropomorphize the animals he depicts, Kunitz instead helps us recognize the impulses we share with them.

The poet's passion for the natural world began early. "As a child, I haunted the woods," he told us, exploring the wild places in his hometown of Worcester, Massachusetts, "partly because they represented to me a kind of wilderness that I think the poetic imagination requires. The danger is of becoming too civilized. One needs to be able to deal with chaos and with losses, and we have to believe that there is an unconscious world that is wild, not to be tamed, and needs to be explored if you are to find out who you really are."

The constant tension between the cultivated and the wild is a powerful refrain in Kunitz's poems. "The Testing Tree" moves from a remembered childhood afternoon in the woods, "where the bees sank sugar wells / in the trunks of the maples," to a recurring dream in which the poet's mother "is wearing an owl's face / and making barking noises." He leads us into worlds we can easily recognize, as well as territory that is entirely new. In *The Wild Braid*, Kunitz describes the challenges of shaping each idea, each phrase in a poem, of making sure that he doesn't "cut away the heart of a poem, and are left only with the most ordered and contained element. . . . You must be very careful not to deprive the poem of its wild origin."

This is a landscape that we need our poets to traverse—the edges of experience, between the known and the unknown, so they can bring back to us visions of what lies beyond our own lives, or sometimes a glimpse of what lives inside us that we can't see. As with his garden path, Kunitz believed that an essential element of any poem is mystery. "If the terrain were familiar, the poem would be dead on birth," Kunitz wrote, adding that "the path of the poem is through the unknown and even the unknowable, toward something for which you can find a language."

"One of my deepest concepts," Kunitz told us, "is that the web of creation is a continuous tissue whose invisible filaments envelop every living organism. All of us are connected, are related. If you touch the web at any point, if you disturb it, the whole web trembles. That's the basis of my theory about all living creatures."

Stanley Kunitz died in the spring of 2006, just shy of his 101st birthday. The low stone that marks where he and his wife are buried is engraved with branches and leaves, and a small hand on which perches a little bird. It is inscribed with the last two lines from Kunitz's poem "The Long Boat":

he loved the earth so much
he wanted to stay forever.

Conceptual artist Mel Chin

Reviving a Landscape Through Art

"The future of sculpture . . . , for me, is in plants."

By the late 1980s, the web of creation was deeply disturbed on 250 acres of Mississippi River wetlands near St. Paul, Minnesota, an area called the Pig's Eye Landfill. For decades, businesses, industries, and local communities had dumped their trash there without consequence; later, the city's Metropolitan Waste Control Commission poured sludge on top of the old garbage. The landfill was high on the state's list of hazardous waste sites, contaminated by heavy metals such as lead, boron, cobalt, aluminum, mercury, and zinc, all of which had leached into the soil.

The extreme toxicity of this desolate landscape is exactly what drew the conceptual artist Mel Chin to the site in 1991 for an artwork he called *Revival Field*. On about 3,600 square feet of the abandoned dump, the skyline of St. Paul hovering far in the distance, Chin created a bull's-eye target sixty feet across, surrounded by a circle of chain-link fence. Dirt paths crisscrossed the circle as if the earth itself were seen through the sights of a giant rifle. In each quadrant of the bull's-eye, Chin planted a garden filled with carefully delineated patches of Alpine pennycress, bladder campion, corn, red fescue, and romaine lettuce. Chin chose these particular plants because he believed they had the potential to be "hyperaccumulators"—plants that could leach toxins out of the soil. It was something scientists had imagined might be possible, but had not yet tested. Mel Chin sought to prove that theory through art.

"You know the film *The Graduate*? I would've been the guy who, instead of saying, 'Plastics—the future's in plastics,' I would've been saying, 'The future of sculpture or art, for me, is in plants.'" Chin isn't a scientist; he studied art at Peabody College in Nashville and is known for conceptual projects that present witty and sometimes subversive art in unexpected contexts. Deeply focused on social and environmental concerns, Chin creates sophisticated commentary on contemporary issues. In the late 1990s, for example, he convinced the producers of the popular television show *Melrose Place* to let him and other artists incorporate provocative artwork into the show's set as props. One scene included a couple between sheets that were imprinted with an intricate pattern of unrolled condoms; FCC regulations at the time forbade showing condoms on television. In another scene, a character orders Chinese food, and the white takeout bags are imprinted with the ideogram for "human rights," a Chinese character that is forbidden in China.

The spark for *Revival Field* was an article Chin read in *The Whole Earth Review* by an expert in psychedelic mushrooms who suggested that the plant datura (the hallucinogenic and potentially poisonous plant that Carlos Casteneda wrote about in his Don Juan books) might be able to pull heavy metals out of contaminated soil.

Chin acknowledged that most of his friends thought he was a little crazy to consider creating an artwork out of the idea, but he was hooked. "I started roaming in the fields of knowledge outside of art, outside of what I loved, and I came across an article about the capacity of plants that perhaps could accumulate or pull up heavy metals or toxins from soil. I said, 'Well, that's the sculpture I need to do.' I saw it as a real simple carving away of toxins to create something that is aesthetic in the end. It's sort of [like] Michelangelo who has his chisel, and his block of marble, and he starts carving in the Carrara marble, and out comes what we see as *David* now. I began to see this as a new sculpture, where the plants could be the chisel, and scientific process could be the hand that guides it.

"But it could be conjoined with an art idea—a conceptual art idea. It was so poetic. I felt it forced me into realizing, with the poetry of this, that it needed extreme, pragmatic, responsible behavior. So I went on to the second part of the phase, which was to search for the scientific reality of this." It was not a simple task. "After about four months of writing people and phone calls and just going to libraries and reading more and more and digging deeper into it," Chin finally reached a landfill expert at Texas A&M who said he didn't think there was anyone studying the idea of using plants to remediate toxic sites. Just as he was about to hang up, though, the scientist remembered a paper by a Dr. Rufus Chaney that sounded something like what the artist was describing.

Chin immediately contacted Dr. Chaney. "I asked him about datura. He said, 'Well, that might get you high, but it won't pull heavy metals. But if you want to pull heavy metals, I think you've called the right person.'"

Rufus Chaney was a bit skeptical at first, according to Chin. "He was a very responsible scientist, and here's an artist calling him up and talking about his concept. And he said, 'It's not science. It's only an idea. In order to make science, we need a replicated field test. You cannot create science without taking the experiment out into the field and doing this.' I said, 'Well, then, let me do it. Let's work together. And I will do it responsibly to find you the data, so you can create your science.' So it was a conceptual artwork that would bring science into being." Chaney agreed to advise Chin on the project. "*Revival Field*, therefore, was the first test in the world of this kind of technology for the scientist."

Chin was rigorous about the scientific aspects of the project. He created a control, measured the contaminants in the soil before he planted, and weighed and measured the plants he harvested. These were then dried and burned to concentrate the ore. The goal was to see if the plants could pull enough heavy metals from the contaminated soil to make the plants themselves valuable, since the metals they contained could potentially be concentrated and reused.

"We did the original test, and results that came out showed there was uptake by [some of] these plants far beyond the other test plants. This indicated to the scientists that this is a possibility, that this is a reality. So the first part of *Revival Field* was to confirm and create a scientific technology. Now we're at the stage where there are two hundred and fifty scientists who say they have developed this idea. Private industry is creating businesses out of it, with the intent of selling the product."

The science of *Revival Field* is sound, and has been replicated. But Chin is often asked, "How is this art?" He has a ready answer. "Within the definition of the art world, I would say it's a piece of artwork. It's a piece of conceptual artwork. That's within our world. Remember, I was reading the language of the scientists, too. In their world, it's a scientific project. It's opening up a new field based on what we have learned from *Revival Field*. So, yeah, that art is a science. Sometimes I tell people it doesn't matter to me anymore. It's more what we have learned from it and what is now created." A drawing from the project is in the permanent collection of the Walker Art Center, and Mel Chin's *Revival Field* has also been noted in science textbooks as one of the first experiments in what is now called phytoremediation—using plants to help clean up contaminated soil.

Landscape architect Julie Bargmann

Exposing the Beauty of Industrial Decay

"My role . . . is to reveal how the landscape is one big machine."

The ideas Mel Chin developed in *Revival Field* have had a far-reaching impact on how industries such as the Ford Motor Company approach the damage that manufacturing has inflicted on the landscape. In 2003, Ford began experimenting with phyto-remediation on a huge scale in its enormous manufacturing facility in Dearborn, Michigan, on the banks of the Rouge River—a river that was once so polluted it caught fire. Today, in addition to touring the automotive production floor, in the warmer months visitors to Ford's River Rouge facility can wander along a wood-chipped path through brownfields contaminated by the toxic by-products of almost a century of steel manufacturing. The land is now covered with apple orchards and ponds lined with plants to clean storm water runoff before it reaches the river. Ford has also planted thousands of native perennials and shrubs to see which have the capacity to pull up the PAH (polyaromatic hydrocarbon) compounds left in the soil by the coke ovens that had been used to make steel for Model T's and Mustangs since 1918.

This newly verdant terrain springs from an idea by landscape architect Julie Bargmann, who is drawn to places that most of us avoid. "I do remember a circumstance where it struck me that my attraction to these landscapes was a tad perverse. The first time I flew over the River Rouge Plant in Dearborn, Michigan, the Ford Motor Plant, I was looking out the window and I said, 'That is so beautiful!' People cranked their heads around and said, 'That's very odd.'"

Bargmann has been acclaimed for her lyrical yet practical approach to developing ways to reclaim these landscapes as places where people can work and play. Weaving art with science, she has designed solutions that both heal the environmental damage and preserve the memory of each site's industrial past. With her company D.I.R.T. (an acronym she's variously described as standing for "Design Innovations Reclaiming Terrain" and "dump it right there"), Bargmann has transformed old navy rail yards into a vibrant public square for Urban Outfitters in Philadelphia and designed a park for the Environmental Protection Agency on the site of a landfill, with plans to turn methane gas into energy to heat a local high school. She's used plants to revitalize wetlands in Antioch, Illinois, and created a strategic plan for remediation and redevelopment of a 14,000-acre bauxite mine in Arkansas.

In 1999, Bargmann and D.I.R.T. were part of a team, led by the architect William McDonough, that was brought in by Henry Ford's great-grandson, William Clay Ford Jr., who charged them with "transforming River Rouge into the model of twenty-first-century sustainable manufacturing." The plant included a facility that manufactured tires; 120 miles of conveyors, coke ovens, and rolling mills; and over ninety-three buildings housing every aspect of car making. Bargmann said that William Clay Ford Jr., "did not want to abandon this beautiful but very much rusting and dated industrial plant, [but] instead wondered how could it be reinvigorated and regenerated."

Ford began to implement the plan in 2002. It's expected to take twenty years to complete. "At first the Ford Motor Company was calling it 'Greening the Rouge,' and it drove me out of my mind. I just said, 'Stop calling it that, because it just sounded like the old 'Oh, we'll just throw in some plants, just add water, and it'll all be green, that's fine.' Versus what my argument was and what I felt my role as a landscape architect was, to reveal how that landscape is one big machine; and that in looking at its processes it could be a healthy landscape or one that is self-destructive, which often industry is. So there's always a supposition that industrial landscapes are destructive, and that's not necessarily the truth. They don't need to be."

In her plan for the auto plant, Bargmann focused on two main areas of the industrial site. "There was, to the south, an abandoned coking plant, and to the north there were the plans for the future assembly building. So it was going to be this great narrative of how we look at industry of the past, and how do we regenerate it. We were looking at the science of bioremediation, which is in this case phytoremediation, plants that basically stabilize or extract contaminants from the soil. . . . The conventional practice is excavation and disposal. In the industry they call it 'hog and haul.' Or 'muck and truck.' Instead, we advocated that they carefully look at the cultural value of these structures. There are incredible stories about great-grandfathers starting there. We could juxtapose that with a new layer of regeneration with bioremediation. We felt like that would be this beautiful juxtaposition of valuing the past—not chucking it away—but also recognizing the by-product it has left us, and that we are quite creative in knowing how to regenerate that."

Bargmann has often been called in to address sites like this. One of her best-known projects is a park in Pennsylvania's coal country, on the site of an old mine in Vintondale, in which she partnered with a hydrologist, a historian, and an artist to revive the site while also telling its story. Before they began, the water bubbling out of the mountains of mine refuse—called "black bony"—was a brilliant orange. This is acid mine drainage, which carries with it minerals and chemicals that leach out of the mines. Pictures taken before Bargmann began work on the project show vast swampy areas stained the color of orange juice spiked with rusty nails. In the 1980s, the site was strewn with the remains of leveled coke ovens and collapsed mines, acres and acres of black bony and bricks and steel. Bargmann and her colleagues designed a series of ponds, employing natural systems that slowly decrease the acidity of the contaminated water until it has a pH level that can sustain plant and animal life in newly created wetlands that cover the land once filled with black debris. Bargmann planted the banks of each pond with trees whose leaves echo the changing color of the water: red maples reflecting in the orange runoff, and blue-green spruces surrounding the final, clear blue pond.

In the wetlands, the brick-and-mortar footprints of the original colliery buildings rise out of grasses and shrubs. A mine entrance is blocked by a huge stone, etched with an image of miners getting ready to descend into it, taken from a frame of a home movie made by one of the workers. This vista of plants and birds and fish also keeps alive a memory of the hard human work that was done on the site, and which helped transform the twentieth century.

That history, and the understanding that this industrial past is fast receding, is powerful for Bargmann. "When workers describe these places, one of the first things they describe is the sound and the smell. I think it is something we're going to, in an odd way, miss. This is very physical—I know my reaction to these places is very visceral." She teaches at the University of Virginia and leads her students on field trips to visit decayed industrial sites, like the Roebling steel factory in New Jersey, just south of Trenton. "I found that often I was taking my students to the empty hulks of the factories, and that that was problematic because in a way they were just seeing them as objects. [To remedy that], I took them down the road to an operating factory, and their jaws dropped. In there was the smoke, the heat, it was almost like looking through the smoke and all of a sudden seeing a human face."

Bargmann studied art at Carnegie Mellon University. "I went to school in Pittsburgh, and I'll never forget, as an artist—and this is at a time when I didn't know I'd be so obsessed with industrial landscapes—I explicitly remember a sculpture instructor bringing us to a steel factory. I remember the heat. It was visceral. What's so interesting about industry is these senses. There are those sensations that I think we're going to lose, and there's something to be valued and remembered, they do trigger something in us."

The bucolic campus of the University of Virginia is a dramatic contrast to the landscapes that Bargmann transforms. She brings that gritty reality into the classroom at the beginning of each course by projecting photos of the refineries and chemical plants that line a stretch of the New Jersey Turnpike, viewed through hazy exhaust. Bargmann lets her students know that this is what she considers a beautiful landscape. She said she can trace her fascination with industry back to trips in the family station wagon with her parents and seven brothers and sisters. "I've been working with industrial landscapes the past fifteen years; it was just a natural gravitation toward them. But it wasn't until the past few years that it dawned on me that I grew up in New Jersey, the Garden State, and I drove past the refineries on the New Jersey Turnpike into [New York] City, where my dad worked at the Chrysler Building. I realized that I took that landscape as *my landscape.* I never considered it 'other.' And that's what I've been finding more and more folks are realizing—that this industrial landscape is what they've grown up in and around, and it's a part of them, for better or for worse."

In college Bargmann thought she would become a sculptor. But after she finished graduate school in fine arts, she entered what she called "my black hole period of not knowing what to do with myself. It was then that I discovered landscape architecture, and I realized that I was very interested in that medium and that environment. I was really tired of making those precious little sculpture objects. And right then, for whatever reason, my mother sent me a photograph of myself at four years old, and I was playing in the dirt. I said, 'Mom, why didn't you send me this earlier? I could've known what to do a lot sooner.' "

Some observers have suggested that Bargmann is still a sculptor, but on an enormous scale. She's flattered by that response to her work. "I think public artists, environmental artists, have recognized that in using the landscape as a medium, it's not just a medium. It's not just the material. It's a social context. It's political context. It's complex context. I became completely fascinated with this complexity and really love mixing it up and having it be complicated, rather than a more isolated circumstance where I was inside my own head reacting to the outside world. I wanted to be in the world and have engineers and scientists and bulldozers, and have all this stuff that put me into this larger context, which I find infinitely fascinating. That's why I get totally charged up and it's an obsession. And I'm never bored—with every single site, you have to learn what processes took place there. I'm becoming this expert in slag. I mean, it's weird, but I love it. Each time you're learning these processes, and you're learning 'What is the next step in this process? What is the next evolution in this landscape? Why are we stopping here?' "

Not everyone sees the world the way Bargmann does, and often the people who control the purse strings may not share her vision. D.I.R.T. has produced countless drawings and schematics for projects that won't ever see light. For all her swagger and optimism, Bargmann admits she also gets "so frustrated in my work, and I bang my head a lot up against [people who] can't wrap their head around really embracing these landscapes. They really look at me like I'm insane. I run up against it a lot with the corporations I deal with, the agencies. I'm up against a particular view."

But Bargmann knows that being involved with landscapes means understanding geologic time. Even though *Time* magazine once named her one of the hundred top innovators to watch, and she has revitalized brownfields into parks ripe with both nature and a sense of history, her most enduring impact may be realized through the students who come to the University of Virginia and take field trips to derelict factories to learn how to heal the land. "That's why I teach, and I do projects to bring those lessons back. A lot of my lessons are about the battles, and so I teach because I like to think there's gonna be an army coming, and they're going to be able to build some pretty amazing things."

Chapter 4

GOING HOME

The May afternoon in 2000 when I met Kurt Andersen for the first time was unusually warm. As I walked toward the diner Kurt had chosen for our meeting on the far west side of Chelsea, in Manhattan, I pondered what I already knew about him: founder of the iconic *Spy* magazine in the 1980s, journalist for *Time* magazine, author of a well-regarded novel. Kurt's smart, funny voice on the page was familiar to me, but I was eager to hear what his speaking voice sounded like. I was contemplating whether to join *Studio 360* as executive producer, and as I began to imagine how the program would sound, I knew that his voice was crucial for the potential success of the show.

When I walked into the diner, I immediately noticed a man with dark hair, stylish glasses, and intelligent, dark eyes. I know this could describe a million or so male New Yorkers, so perhaps I should also note that Kurt is very tall, too. As we made small talk, I listened to his infectious laugh (a great asset for someone who wants to succeed in radio), and, to my surprise and delight, the subtle but unmistakable trace of a midwestern accent. Kurt's wry, ironic style as a writer had struck me as quintessentially

New York, but his speaking voice reveals his roots in Omaha, Nebraska.

While many people (including, I later learned, some of the vocal coaches Kurt had worked with as he prepared for his role as host before I arrived) think that radio hosts should have perfect, untraceable diction, I strongly believe that to be a great radio personality you should sound like yourself, and like where you come from. So much of what communicates on radio is not the words that are being said, but the tone and quality of the voice that says them. Our initial response to the people we hear on the radio is, inevitably, emotional, and an authentic voice will always elicit a more visceral response than a generic, perfect voice. Kurt's voice, with its Omaha inflections, is wonderfully authentic.

Kurt told me recently that "on a certain level I think of myself as someone from Omaha who has lived in New York for a long time, rather than a New Yorker." He grew up on Ninetieth Street, which was then the western border of Omaha. "On Ninety-sixth Street when I was a kid," he said, "cornfields started. So it was really this sense of being out on the edge. Now that's the center of town."

His mother had fallen in love with their house years before his family purchased it, when Kurt was three. Rambling and idiosyncratic, it was "full of little nooks and crannies to crawl into." As Kurt described the house he grew up in, I heard great affection in his voice, and an inkling that he had honed his keen sense of observation at home, early on. "The way this house was designed, there were rooms that jutted out by themselves and thus got light from three sides," he told me. "I remember sitting and staring at the afternoon light, the sunshine coming through windows, and, literally, staring at dust motes for I don't know how long, as they played through light."

Kurt's house sat in the center of an acre of land, complete with a lily pond, a greenhouse, and a tennis court, all a bit run-down. “It makes it sound like some super-duper mansion, which it wasn’t. In 1920, when [the first owners] built it, they had enough money and ambition to build this stuff around the house.” More exciting for Kurt and his older brother and two sisters was the acre of woods next door. “We called it the Vacant Lot—capital V, capital L. It’s funny, when you realize, ‘Oh no, vacant lot is just a generic thing—not *the* Vacant Lot.’ Similarly there was a creek right near our house called the Papillon, but we called it the Pappio. That was its nickname in general, in Omaha, and I remember at age six realizing that Pappio was not a generic name for creeks. My mother always said she let me believe that for years longer than I should’ve. She found it so cute that I called all running bodies of water ‘pappios.’ ”

Kurt and his siblings spent hours in the Vacant Lot. “It had its distinct botanical character, a full range of topography and geography and kinds of trees and plants that were different from place to place. It was sufficiently overgrown that if you’re a little kid you can really lose yourself and imagine that you’re fighting the Nazis in the Ardennes. It was a fantastic place. One of the great bummers of my young life was when somebody bought it and began building condos there. It was like the official end of my childhood.”

At eighteen, Kurt left Omaha for Harvard, and after graduation moved to New York, where he’s lived ever since. But he would often go back to visit his parents, and over the years working together on *Studio 360*, Kurt and I talked about sending him to Omaha to explore the burgeoning arts scene there. In 2004, we finally had the chance, and Kurt found a vibrant community of musicians and artists in his hometown. He talked with the indie band Tilly and the Wall, which boasts a tap dancer rhythm section, toured the new performing arts center, and loved visiting the collection of artists’ studios in old warehouses downtown. “When I went back and did that piece for the show,” Kurt said, “it was so interesting to me that it would be less unthinkable today for the eighteen-year-old me to think, 'Maybe I'll stay there.' I now encourage young artists and people I know all the time, if they can't live in New York, to go live in Omaha because it's so cheap and a nice place to live with other young artsy types."

Kurt has never put Omaha in his fiction, although he's come close. He told me that the novel he's working on now includes protagonists who grew up in the 1960s, like he did. "For certain narrative reasons I needed them to be near a big city," Kurt said, and he had originally imagined that they might grow up in New Jersey, where I live. "But then I realized, while I can imagine anything and get the details right and it'd probably be fine, I would know what it's like to live in Wilmette, Illinois, in 1963 more than I would get what it was like to live in Maplewood, New Jersey, in 1963. So 'write what you know' is not an absolute rule, but writing novels is hard enough without setting yourself new challenges of understanding what it was like to be four miles from Newark during the riots of 1968. Wilmette is not Omaha, but it is a certain kind of midwesternism of a particular era."

In *Studio 360*, many artists have talked with us about drawing deep inspiration from where they grew up. For almost fifty years, the photographer William Christenberry has traveled back to Alabama to photograph the farm buildings, churches, and roadside cafés in Hale County, where he spent summers as a child on his grandparents' farms. The Oscar-winning director Alexander Payne hails, like Kurt, from Omaha. To make many of his movies, Payne has returned to Omaha time and again, seeking to capture a sense of the Midwest he's never seen on film. And the writer Chimamanda Ngozi Adichie grew up on the campus of the University of Nigeria in Nsukka during the 1970s and '80s. In her award-winning second novel, *Half of a Yellow Sun*, she imagines what life was like in Nsukka during the bitter civil war in Nigeria, a conflict that took place less than a decade before she was born, yet few people discussed it while she was a child.

None of these artists still lives in the place where they were born, yet for each, their imagination returns again and again to where it was formed. Their journeys home offer us intimate glimpses of a beloved place, conjured with the vivid details of memory and tempered by the grit of contemporary reality.

Photographer William Christenberry

The Landscape of Childhood

"My work . . . [is] going back to the same places
time after time and watching them
decay or change or disappear."

A solitary one-room building stands at the edge of a pine forest in one of William Christenberry's best-known photographs. Red bricks cover the structure's entire façade, from the floorboards to the peaked roof. Four rickety wooden stairs lead up to the front door, which is also covered in bricks. In this picture, the building takes up almost the whole frame, and like a child's memory of home, it dwarfs everything around it. Another photograph captures the structure from maybe a hundred feet away across a grass clearing, and here you can see just how tiny the building really is, nestled among the towering pines.

William Christenberry calls it "Red Building in Forest," and for more than thirty years, he has made an annual trip back to Hale County, Alabama, to take its picture. "I like to take back roads off the main roads and just see what I might stumble upon.

In 1974, to be exact, I came across this little building. For many years I did not know its function, its purpose. It was just there. Nothing to put it in context. Nothing nearby, man-made. Until one year in the mid-eighties I was there, and here was this metal sign on a steel post in front. It said Beat 13. That was my clue. It was a polling place. It was a place where people in the backcountry could come and cast their vote on voting day rather than going fifteen miles down the road to the county seat. Before that, a long time ago, it was a one-room schoolhouse."

Christenberry's photographs of the "Red Building in Forest" are part of his decades-long study of Alabama cafés, barns, homes, and churches. His pictures always hint at the rich stories these buildings might tell if they could speak. Since the late 1960s, he's made an annual pilgrimage to Hale County to capture the rural landscape: storm clouds above a pear tree whose fruit hangs over a dilapidated fence; kudzu engulfing an old farm building; rust-eaten and bullet-ridden roadside signs; white churches at the end of long dirt roads bordered by tall trees. "By the time I get there every summer, and it's usually summer, there are pictures," he said. "I'm chomping at the bit to get out in the countryside and see what else I can find . . . often going back to the same places time after time and watching them decay or change or disappear."

Christenberry was born in Tuscaloosa, Alabama, in 1936. "That's where my heart still is. I have nothing but fond memories of my growing up in Alabama. Tuscaloosa is a university town, a fairly large city, but my grandparents on both sides, the Christenberry and the Smith families, had farms just south of there, in Hale County. I used to go there in the summer, spend two weeks at the Christenberry family home, and two weeks at the Smith family. I became infatuated with that landscape and the things in that landscape, especially the vernacular architecture, and many of the traditions that are still there."

Christenberry's images of Hale County unfold like time-lapse photography, not of a flower opening over several days, but of a place transforming over many years. One powerful series of images captures Coleman's Café, in Greensboro, Alabama. Christenberry first shot the wooden, slightly decrepit façade of the roadside building in 1967, with its round Coca-Cola sign above a rectangular white sign proclaiming COLEMAN CAFE in big, capital letters. Ten years later, the Coca-Cola sign is rectangular, the two screen doors are covered in torn brown paper, and the front wall is painted brick red, a striking contrast against the dark-green trees and the deep blue sky. Later, we watch as the building crumbles, year by year, board by board, until all that is left in 1996 is a pile of broken wood by the side of the road.

As a child, Christenberry loved to draw, and, later, while a student at the University of Alabama, he made large, brightly colored expressionistic canvases of tenant houses and graveyards. "I always said those paintings look like Soutine and de Kooning in the same painting." For guidance and inspiration, he took snapshots of his environment with a Brownie camera he'd been given as a Christmas gift.

In 1961, after receiving his MFA from the University of Alabama, Christenberry decided to try his luck as an artist in New York City. "It was the year that Bill Christenberry had to grow up. I had eight different jobs in twelve months; remember, I had my master's degree in painting, but I didn't want to teach. I did all kinds of things"—such as selling men's clothes in Greenwich Village, and cleaning Norman Vincent Peale's church on Fifth Avenue. "I ended up being a file clerk in the picture collection at Time Life. And Walker Evans saw fit to help me get that position."

Christenberry clearly remembers his first encounter with Walker Evans's powerful photographs. "I was at a bookstore in Birmingham in 1960, and there on the counter, near the cash register, fresh from the publishing house in Boston—Houghton Mifflin—was this book *Let Us Now Praise Famous Men* by James Agee and Walker Evans. I started thumbing through the first pages, and, as you know, in that book the photographs and Agee's writing are coequal. The photographs are not illustrations. [It's] one of the best examples of collaboration between a visual artist and a literary artist.

"I realized that I knew some of the people in Walker Evans's photographs. They did not sharecrop from my grandparents, my mother's family, the Smith family, but they lived nearby. And they would come over to my grandparents' house and help my grandmother do washing and things like that. A poor, white family." The 1960 reissue of Evans and Agee's groundbreaking 1941 book was a revelation for the young painter. Christenberry found out that Evans was a senior editor at *Fortune* magazine, and after arriving in New York, he got up the nerve to look Evans up. They became friends, and, Christenberry recalled, "he and Mrs. Evans were very kind and supportive of me."

One day, Evans asked to see Christenberry's paintings. He and his wife "came down to my little place on Seventh and Leroy Street. Walker said, 'You mentioned one time in our conversation that you made color snapshots as a reference for these.' I said, 'Yes, sir.' And he said, 'I'd like to see those.'

"Man, I was intimidated—well, that's not the right word—I didn't know what to say. I was pleased, but also nervous. So he and Mrs. Evans came back another time. I had sixty-one of these little Brownie snapshots processed at the local drugstore, and I dry-matted [the photos] on a piece of mat board a little larger than the print, which was fortuitous—it gave them some backbone, some substance, and they didn't curl up."

Evans's evaluation of those snapshots had a dramatic effect on Christenberry. "Mrs. Evans and I were over there, talking, and he just distanced himself completely from the conversation. I didn't think he would ever finish looking at them. Finally, he turned to me and he said, 'Young man, there's something about the way you use this little camera. It's become a perfect extension of your eye, and I suggest you take these seriously.' I was about as interested in photography as I was in physics. Zero at that time. But I was photographing things that I knew, that I loved, that I empathized with, and nothing more." The young artist listened carefully to Evans's critique, and began focusing more and more on taking pictures.

In 1962, Christenberry left New York and moved south again, to become an assistant professor at Memphis State University in Tennessee. A few years later, in 1968, he was hired as a painting teacher at the Corcoran School of Art, and he and his family moved to Washington, D.C., where he's lived ever since. The artist is often asked why he chose not to live in Alabama, which has so clearly captured his heart and imagination. But he believes it's essential that he left. "My feeling about that is that if I lived there, I don't know if I would see it as clearly."

The Corcoran was the first gallery to exhibit Christenberry's snapshots, in 1973. Walker Evans heard about the exhibition, and Christenberry said he called him on the phone one day to ask about it. "He said, 'I understand you're having this exhibition—is there going to be a catalog?' I said, 'No, sir, but there's going to be a very nice brochure. A color foldout of pictures.' He said, 'I'd be hurt if I'm not asked to write something.' Well, I passed out when he said that. Of course, I wasn't going to ever ask him. He wrote about the innocent eye—not the naïve eye, but the innocent eye—in the best sense of that word, 'innocent.' And he summed it up: the last line is, 'And each one is a poem.' I can't expect any higher praise. I see my photography, I see all of my work, as an attempt to create some kind of visual portrait."

Each of Christenberry's pictures has a protagonist, yet people are rarely glimpsed in them. Instead, it's what people have made, what hands have built, that he portrays: the symmetrical twin steeples of a bright white church in Sprott, Alabama; a wide warehouse in Newbern with a roofline that resembles a witch's hat, painted the same faded green as the parched grass in front of it; fluffy white clouds in a blue sky suspended over kudzu that grows like leafy fur on a building in Greensboro, slowly devouring it. The colors are vivid, the structures centered in the frame like a face, the atmosphere elegiac.

These are iconic images of the Black Belt agricultural heart of Alabama, yet Christenberry's photographs consistently avoid cliché. His pictures are as much about the washed-out turquoise paint on an old car and the slanted southern light hitting a sun-bleached red roof as they are studies of the Hale County landscape.

Christenberry has also created sculptures of some of the buildings he loves. With these he captures in three dimensions his memory of a specific place, and sets each structure, made of wood and metal and paint, on a bed of the red Alabama soil that he collects on his trips. Along the way he's also convinced farmers and shopkeepers to give him the signs that he loves to photograph. "Most of the time I would ask the store owner, if there was someone there, 'Would you part with that, or what would you take for that?' Invariably they say, 'You like that old thing? Just take it, son.' So I have a wonderful, I don't want to call it a collection, but assembling of these things over time. It's what I call the aesthetics of the aging process. How time and the elements will affect a Coca-Cola sign, or a Top Snuff sign, or things like that." The walls in Christenberry's light-filled Washington, D.C., studio are covered with these signs, and he will often exhibit photographs, sculptures, and the roadside signs all together, offering the viewer a glimpse of his journey from inspiration to theme and variation.

Not everyone loves what Christenberry has chosen to represent. "Some people think what I do with the camera and my sculpture and drawing is being critical of Alabama. That's not true. The large portion of my work is celebratory. But some people say, 'Why does he like that old junk? Why doesn't he photograph some of those antebellum mansions?' They're too pristine. They're too, I guess in a way, perfect."

Christenberry finds beauty in the everyday buildings of Alabama. Yet he is not afraid to also address the darkness within this landscape, and he has worked for many years on a series about the Ku Klux Klan. His first encounter with the Klan was when he was still living in Tuscaloosa and convinced a friend to go with him to witness a nighttime Klan meeting at the Tuscaloosa courthouse. Because there weren't many people around, he was able to wander inside the building, which seemed to be empty, though the lights were blazing. On the third floor, he encountered a Klansman in full robe and hood. "When I got to the top of the steps, he did not turn his head or his body. He turned his eyes to look at me. I have never seen anything more frightening than those eyes glaring through those eyehole slits. I stopped dead in my tracks and didn't go any further. I went right back down those steps and out of the building."

He immediately began a series of drawings to convey the terror and revulsion he had felt. Next, he began making sculptures, dressing Barbie dolls in handmade satin Klan robes. A few years later, when the company that made G.I. Joe came out with new dolls with articulated limbs, he bought twenty, thinking they might make better Klansmen than the Barbies. "The young lady at the cash register could not contain her curiosity. She said, 'Mister, it is nowhere near Christmastime. May I ask you what you are going to do with twenty G.I. Joe dolls?' I said, 'Young lady, if I told you, you wouldn't believe me.'" Christenberry designed robes for them, which were sewn by friends, and he dressed each doll to create a frightening assembly of tiny G.I. Joe Klansmen.

By the time he and his family moved to Washington, D.C., he had constructed a Klan tableau of some 200 dolls.

Christenberry said he took part in the 1968 Memphis Sanitation Workers' strike just before Dr. Martin Luther King was killed. But he admitted, with some regret, that he's never been much of a marcher or a joiner. He's chosen to wrestle with the legacy of racism in his home state by creating a visceral experience with this work. "For a long time it was not seen because it was private work, but when it was made visual, the response was: some antagonistic, some supportive. And then, lo and behold, from my studio in Washington, right there in a nice part of town, the work was stolen. And it has never resurfaced."

This was no vandalism, Christenberry said. "The door was locked back up. And it was a room off my studio work area where I had put this body of work together—or it was coming together. I didn't go in the room at all unless I was adding a piece or I was getting something out of a closet. On January 1, 1979, I opened the door, and I could not believe my eyes, because the work was gone, just as neat as a pin. I could've better understood if it had been vandalized, things strewn about, broken. It was obvious that somebody really took care. And what really frightens me is it's probably someone I know, because they knew my habits. They knew I wouldn't be there on January the first. I say 'they'; it could've been one person—I suspect it was more than one. My great fear to this day, as I said, [is that] it's someone that I know, and I don't think it was the Klan."

The theft scared Christenberry, as well as his wife and young children. But he continued to pursue the subject, and his work was championed by a daring curator named Walter Hopps, who had been the first to exhibit Christenberry's photographs at the Corcoran. Hopps put together an exhibition of the sculptures and drawings at Rice University in Houston in 1982. The work has been exhibited a couple of times since, displayed in a small, claustrophobic room crammed full of satin-robed dolls on horseback, behind bars, in coffins. Some are bound with barbed wire, others covered in melted wax. Almost every inch of wall space holds drawings of Klansmen, metal relief sculptures of guns, a neon cross. Christenberry calls it *The Klan Room*.

"It's a tough piece," Christenberry said. "It's meant to be. You're not meant to be comfortable in there." The work continues to be controversial, with some arguing that it glorifies the Klan; others, such as *Washington Post* reporter Teresa Wilz, are struck "not by the banality of evil, but by the silliness of it." Christenberry's response is that his intention is to provoke discussion. "I've been criticized for even undertaking this. But my argument is, how can I turn a blind eye to racial prejudice and injustice?"

Shortly after the initial work was stolen, Christenberry had a dream in which he saw a tall white building with an impossibly steep pitched roof, set on a backcountry road in Alabama. This led him to begin a series he calls *Dream Buildings*, stark white needle-like structures with elongated spires. Unlike his other sculptures, in which he evokes buildings that are palpably familiar through his photographs, these towers come directly from his imagination. "Only later did I realize how much those pieces resembled a hooded Klansman." Yet they, too, fit within the landscape of his remembered Alabama, a place he has lived outside of now for most of his life.

William Christenberry visited *Studio 360* in the early summer of 2006, when the Smithsonian was presenting a survey of his photographs, paintings, and sculpture, and he had a big gallery show in New York. While all this attention was being paid to him in big cities, he was busy planning his next trip down to Hale County. "I've made that trek every year, annually, without a miss since 1968," he told us. "For a long time it was with children 'cause I'd take them to see their grandparents. They're grown now, and I make the trek by myself. It's approximately 806 miles to Tuscaloosa, and down into Hale County add another fifteen or twenty. I must admit, this year I said, 'Do I want to change tactics?' It's so hard to fly, to take that equipment, especially through all that security. I have that heavy tripod. But I'll make it sometime next month, in late August. I'm adamant about keeping the string intact because if everything goes well, and it's still there, I will make the annual picture of the 'Red Building in Forest.'"

Director Alexander Payne

Creating a Sense of Place

"I was hungry to see some version of
midwestern life on screen."

Kurt has said that Alexander Payne pretty much nailed the middle-American blandness of his hometown of Omaha in the 2002 movie *About Schmidt.* When Payne visited *Studio 360*, shortly after the movie premiered, he admitted that "there is a magnificent plainness to a lot of Omaha," but added, "That doesn't mean you don't love it any less."

About Schmidt was Alexander Payne's third movie—and the third he'd set in Omaha, even though the book it was based upon took place in New York City and the Hamptons. Why Omaha? "A flip answer that I give, because I've been asked that question a lot," Payne said, "is would you ask that same question of Spike Lee and Martin Scorsese and Woody Allen, 'Why do you shoot in New York?' Or Quentin Tarantino, 'Why does it occur to you to shoot in L.A.?'

"That's where they're born and raised. I am born and raised in Omaha, and it occurred to me, as it does to many writers and artists, certainly early in their careers, at least, to want to explore and express something that's related to where they're from, and where they grew up, and where they have their earliest associations. They either want to explore or express or exorcise something. I'm a filmmaker, and I'm from Omaha, and early in my career I've wanted to capture something about Omaha."

Payne also wanted to portray facets of the Midwest that he hadn't ever seen onscreen. "I'm thinking, too, about 'Midwest'—what does that word mean? It means everything and it means nothing. In American movies we've so often seen those beautifully sunlit cornfields and a woman coming out of the farmhouse and looking at the cornfields, and the man comes in on the tractor. I'm making this up, so it's a little silly, but you know what I'm talking about. But how about the Midwest—how about a drive-by shooting at Twenty-fourth and Lake in Omaha?"

There are no drive-by shootings in *About Schmidt*, yet the movie does depict aspects of the city that may not be familiar to anyone who isn't a native. The film begins with a shot of the skyline in the distance, and quickly zooms in to focus on a skyscraper with the word WOODMEN emblazoned in huge letters across the top. Once the tallest building in Omaha, the Woodmen of the World Insurance tower is a familiar city landmark. It's also where Warren Schmidt (played by Jack Nicholson) waits out the last few minutes of his last day of work. "Jack plays an actuary who retires from Woodmen of the World Insurance Company," Payne explained. "Upon retirement he feels no sense of pride and satisfaction; rather he feels quite bereft. And alienated and in crisis and very depressed. The film is a comedy about the weeks that follow his depression."

It's an unlikely comedy, but Payne finds the humor in Schmidt's travails, and creates a sense of place through the tiny details that clue us in to the location: the framed map of Nebraska on the wall above Schmidt's desk in his house, the photos of wary cattle in the restaurant where his farewell dinner is held, the open midwestern vowels and dropped "G"'s in the condolences his friends offer when they say they're "prayin' for you" after his wife dies. "I'm hungry for art and movies that are related to place, and by extension related to our actual lives," Payne said. Omaha for him is evoked in the smallest of elements, like "the sound of a storm door closing. You hear it two or three times in *Schmidt*, and every time I hear that storm door closing, I remember my own childhood and the sound of spring action storm doors closing. It could be something as tiny and as insignificant as that, but it has meaning."

Even the mercurial Omaha weather has a key role. Early in the movie, as Schmidt and his wife leave his retirement dinner, rain pelts the windshield, and throughout the film the sky is frequently overcast and gloomy. It's a familiar mood for residents of Omaha; Payne said the storms make it challenging to shoot a movie there. "But later when you see that weather on-screen, it's so nice. We are always so happy, the director of photography and the production designer and I, when we wake up and it's overcast or it's windy, or we don't know quite what the weather will be. Unless it's going to present a continuity problem. Just the weather, and to see wind on-screen, and the trees really blowing while people are talking about something completely different in front of it. It's just nice."

Aside from the weather, Payne said he's found it easy to shoot his films in Omaha. "Because you don't get caught in huge traffic jams. Even as immense as Omaha is now, you can still basically get across town in about twenty-five minutes if you know how to do it. And because I'm the only film guy shooting there, locations and police permits and cooperation with the city and cooperation with the location owners—it's really easy there. It wasn't easy at first. I know when I was shooting *Citizen Ruth* and [production designer] Jane Stewart and I would knock on doors, as we still do, to get locations. A couple times shooting *Citizen Ruth*, it was so outlandish that someone was shooting a movie in Omaha that they called the police on us. It must've been some ruse, some criminal ruse. But now, after *Election* and while we were making *About Schmidt*, and it was in the press that Jack Nicholson was coming, we'd knock on doors and say, 'Hi, we're considering this house.' Then it was, 'Please come in and have some coffee.' "

For *About Schmidt*, Payne found a solid brick two-story house in Omaha, which, with its dated floral wallpaper, sloped front lawn, and constant drone of a lawnmower in the distance, could exist just about anywhere in suburban U.S.A. Schmidt's wife, Helen, chops up a chicken on a butcher-block counter in their slightly untidy kitchen, while Schmidt sorts the mail at a big wooden desk in a room nearby. When he drives to the post office to send a letter, the city streets are lined with Dairy Queens and hair salons.

Payne is not happy about the generic changes in his hometown. "It's like this whole country's becoming those backgrounds on *The Flintstones* and those Hanna-Barbera cartoons," he said. "Whatever street you go down in this country, it's Costco, Burger King, Starbucks, it's really, really awful. Even in a place like Omaha. Combine that with the city fathers, who never think twice about knocking down old buildings. I'm not going to deny that a city has vibrancy, and that it's an organism that's constantly evolving and changing. That's true, but architecture is linked in a very, very strong way to a sense of heritage. For example, the Old Market in downtown Omaha is the number one tourist attraction in the state of Nebraska." Artists and musicians have lived and worked for decades in the warehouses that were crucial in Omaha's early life as a crossroads for goods that traveled the country by river and rail. "It's a tiny island of old buildings remaining in a sea of knocked-down architectural heritage. We forget our past so quickly, but we need it for a sense of identity. For some idea of who we are."

In his early films, Payne displays an easy intimacy with Omaha's varied neighborhoods. For *Election*, he follows a couple of high school kids as they drive through suburbs so new that the brick mansions are still being constructed; the title character in *Citizen Ruth* wanders up a rough, hilly street to peer at her estranged children through the broken screen door of her brother's house. "I'm able to make those stories more personal to me by setting them in Omaha, because I want to get the sense of milieu correct. Fellini was asked many times in his career to come to America to make films. On one occasion he said, 'Well, I don't know how they drive there, I don't know how they smoke their cigarettes there, I don't know on what side they wear down their shoes when they walk. I know how they do it in Rome, and I think I need to feel comfortable that I'm getting the details right.' I would [also] have that process no matter where I'm from. I just happen to have been born in Omaha."

Omaha is the principal location for *About Schmidt*, but the movie is also an unlikely on-the-road story. About halfway through the film, Warren Schmidt leaves Omaha to drive through the flat plains of Nebraska in order to reach Denver for his daughter's wedding. The filmmaker was thrilled that the story called for heading out of town. "That gave me a chance, finally, to drive around the state of Nebraska and get to know it a bit more. Omaha is scooched up against the eastern border, and very few Omahaans really know the state of Nebraska all that well. It's a little bit like New York and the rest of the country. It's like Omaha and the rest of the state."

Payne is pleased with the midwestern verisimilitude he's achieved in his first films, and said that a journalist once paid him what he regards as the ultimate compliment regarding *About Schmidt*. "He was talking about one point of the film, and he said, 'That one point in the film gave me the impression that I was watching a well-shot documentary.' It gave me the language to explain to other, future, collaborators the aesthetic that I may be looking for, and that can transcend genre. I can be making a Western or a science fiction movie, it doesn't really matter, but if you tell people: 'Here's the story and here's the setting, and we're going to make a well-shot documentary about this story that we're doing there,' that's kind of what I'm looking for. I feel somehow prepared to do so, having made these first three films in Omaha."

With all his deep affection for the town he grew up in, when Payne visited *Studio 360* in 2002, he was on the verge of change. "I'm actually anxious to leave Omaha for a while," he told us. "I still have an apartment there, but in terms of shooting—on the one hand, I love Omaha and I've enjoyed shooting there, and I think *About Schmidt* finally begins to capture something elusive, which is a sense of place. But I also like language and culture in film, and traveling is a wonderful magic carpet you can use to do that. I want to move on and explore other places. I feel like I've had a good grounding in trying to capture reality by having worked in Omaha."

Just after his visit to *Studio 360*, Payne ventured beyond Nebraska's borders for another road trip, adapting a novel about two middle-aged guys as they take a journey through sunny California wine country, in search, as the movie's tagline proclaimed, "of wine. In search of women. In search of themselves." It was called *Sideways*—and with it Payne won his first Oscar.

Novelist Chimamanda Ngozi Adichie

Reviving the Past

"I grew up in the shadow of the war."

At first, we see the streets of Nsukka, Nigeria, through the eyes of Ugwu, a thirteen-year-old boy, in Chimamanda Ngozi Adichie's luminous and devastating novel *Half of a Yellow Sun*. Ugwu has traveled miles from his home village to become a houseboy for a professor at the University of Nsukka, and as he walks into town, he savors the sensations of this unfamiliar landscape, noting the heat of the street's surface through the thin soles of his shoes, the delicious scent of the flowers that flank the entrance to his new master's home, the butterflies fluttering above the green lawn in the sunlight. On the doorstep, he wants to touch the cement walls of the house before he enters, but doesn't; he's curious about how different they might feel from the mud walls of his mother's hut that still bore the imprint of her fingers. Once inside, Ugwu is overcome by wonder that he will "sit on these sofas, polish this slippery-smooth floor, wash these gauzy curtains."

Chimamanda Ngozi Adichie grew up in a house very much like this one. "The floors were always very cool, which felt quite nice; they were painted a deep burgundy. It's the kind of house that one might expect British academics to put up when they're setting up a university in the tropical part of the world." Adichie's family moved into the two-story house in 1982, when Adichie was five years old, after her father, a professor of mathematics, was named deputy vice chancellor of the University of Nigeria. Located on Marguerite Cartwright Avenue in a fancy part of the campus, the house was set on a wide lawn with mango and pear and cashew trees. "It had five rooms and a study, and a large living room, and a large dining area as well. The staircase was quite grand and high, and behind it was a little storeroom where we stored all sorts of cottons and things, and that led to the kitchen, which was quite large. The most striking thing about it was the staircase," which was so imposing that on her first night in the house, Adichie was terrified to walk upstairs to the bedroom she was to share with her siblings. Within a few days, she was sliding down the banister on pillows with the rest of the children.

The house on Marguerite Cartwright Avenue was where, at the age of seven, Adichie first wrote little stories for her mother. Three years later, she sat at her father's dusty, imposing desk in an upstairs office to write her first "book." "I imagined I was a writer when I was fifteen, because I was writing bad poetry," she said. "Even then, I was very dedicated. I would write these poems. I would find the addresses of these magazines and send them off. I would ask my mother to make copies of things for me when she went to work." Her work was published in Nigerian magazines before she had graduated from high school.

When she began her university studies, Adichie took the classes she would need to get into medical school. "I was expected to become a doctor because I did well in school. And this was a general expectation of people who do well in school. You become a doctor, or an engineer, if you were a boy, or a lawyer, at the very least. So I went along with that even though all I ever really wanted to do was write." At nineteen, with her parents' blessing, she quit her medical studies and left their house and Nsukka for the United States, where she attended Eastern Connecticut State College. "My entire life, until I left Nigeria, was in that house. And when I went back to Nigeria after graduating from university in the U.S., it was back to that house. I just really loved that house."

That house has a literary history, too. Before the Adichie family moved in, it was the home of the Nigerian novelist Chinua Achebe, author of the seminal 1958 novel *Things Fall Apart*, about colonial life in his country. "That Chinua Achebe had lived in that house seemed fairly ordinary," Adichie said. "People live in houses before others lived in houses, this is the way life works on the campus. It wasn't until just before my book was published and I told my editor about it, and I said to her, 'It's interesting that I lived in this house,' and she said, 'This is very, very important.' Only then did I start to realize that it is, in fact, very, very important that I lived in this house previously occupied by Chinua and his family." Adichie laughed as she continued, "Only then did I start to think about the stories I could tell to advance my career about how Chinua Achebe's literary spirits had whispered to me at night when I got up to use the bathroom.

"Of course, at this time Chinua Achebe had become a very important writer to me, personally. I had come to deeply love and value his work. I grew up reading a lot of British books for children and didn't realize that people like me, little Igbo girls who are from eastern Nigeria, could exist in literature, in fiction, until I read *Things Fall Apart*. I really do think that it changed the way I looked at what literature could do and be. And it wasn't an automatic thing that happened overnight. Slowly, I moved away from thinking about literature and writing stories that were about white people in England. Because it made me see that people like me, that stories that I found familiar, could exist in fiction. I often like to say that Chinua Achebe gave me permission to write my own stories. And I do mean that. *Things Fall Apart* started my journey towards the self-awareness that I could write the stories of the people who I knew and saw."

These people fill the pages of *Half of a Yellow Sun*. Like Adichie's own home, the house that Ugwu enters on Odim Street in Nsukka contains and amplifies the complex lives that play out between its walls. Ugwu works for Odenigbo, a charismatic mathematics professor who teaches at the university and plays tennis at the faculty club. The refrigerator is always full of food, the living room shelves overflow with books, there's a gleaming white tub in the bathroom. Most evenings, an eclectic group of academic colleagues gather in Odenigbo's living room for enthusiastic arguments about the disasters of colonialism and the future of the Igbo people in Nigeria. It's the early 1960s, the university is only a few years old, and Nigeria itself has just climbed out from under sixty years of British colonial rule.

Odenigbo's lover, Olanna, a beautiful sociology instructor, soon joins the household, and through her Adichie shows other facets of life in Nigeria. Olanna visits her wealthy parents in their ten-bedroom house in Lagos, and she later travels to the north to spend time with her aunt, uncle, and cousin in a village outside of Kano. Even as she studiously avoids glancing at the black cockroach eggs that lodge in the corners of the kitchen table near where her aunt cooks a chicken that not minutes before had strutted around the yard outside, Olanna feels more emotionally at home in her aunt's communal household than she does in the icy abundance of her parents' lives of fashion and fancy parties.

Through these very different homes, in a university town, a big city, and a country village, Adichie creates a resonant picture of the vibrant life in this newly independent African country. But the buoyant mood lasts just a brief time; soon, the highlife music that Olanna loves to listen to on the radio is replaced by news reports of military coups. Her aunt, uncle, and cousin are slaughtered in a massacre of Igbo families in the north; she finds their mutilated bodies bleeding in the dust. It's the beginning of the struggle that leads to the declaration of the independent state of Biafra, and to civil war.

Adichie was born in 1977. "I remember being taught 'Oh, a civil war happened; it ended in 1970; Nigeria was united and all was well.' And I have a lot of friends, and I think generally people of my generation, born after the war, we just have no clear sense of what it was about."

Perhaps because so little was discussed, Adichie found herself obsessed by the war. Before *Half of a Yellow Sun*, she wrote many stories about it, "most of them very bad. . . . I was trying to get to a place where I felt confidence enough to write long fiction about the war and to deal with many characters at the same time. In some ways I think I had a romantic fascination. I'm still fascinated with that time, but it's no longer as romanticized as it was. I started to read quite a bit about that period before the war, [when] the colonial government achieved independence. Because I think that it's almost impossible to think about the war, to understand it, without understanding what happened in the years before . . . I was just waiting to be ready emotionally to write the book."

One can understand why Adichie needed time to prepare to write *Half of a Yellow Sun*. The novel contains many interwoven strands: Ugwu's coming of age, the story of Olanna and her "revolutionary lover," as her twin sister, Kainene, calls Odenigbo; and the adventures of Richard, a British writer who falls in love with ancient Igbo brass pots, as well as with Kainene. Adichie weaves the bright threads of their lives through the dark fabric of a country at war.

When the Nigerian army invades Nsukka, Ugwu and the family he serves—Olanna, Odenigbo, their small daughter—are forced to flee their comfortable home. Their lives undergo a series of diminishments, as they move to other towns and villages and ever smaller homes in order to escape the fighting. Toward the end of the war, the only shelter they can find is a room with no electricity in a compound with nine other families. They must use a communal kitchen at one end of the building, and communal bathroom at the other. When they arrive, "Olanna looked at it and could not imagine how she would *live* here with Odenigbo and Baby and Ugwu, eat and dress and make love in a single room."

The comfortable life they first enjoyed, complete with a lovely house, meat for dinner every night, and an atmosphere of good-natured argument, is so familiar to middle-class readers anywhere that their losses are all the more harrowing. We empathize with Olanna's despair when she can't find enough food to feed her daughter, and her terror when soldiers invade her neighborhood and drop bombs on the local elementary school.

Both of Adichie's grandfathers died in refugee camps during the war. "In some ways, I grew up in the shadow of the war. I heard stories about what my parents had gone through, little stories. I didn't grow up knowing the big, important things about the war. I grew up knowing that my mother had lost her wig, which had been bought in London in 1964. And that the reason I did not know my grandfathers was because of this war."

In *Half of a Yellow Sun*, Olanna, Odenigbo, their daughter, and Ugwu survive, and when the war ends, they return to Nsukka. Their books have been burned and left in a charred pile in the front yard, the grass in back has grown tall as a man, the bathtub is filled with rock-hard feces. They clean it all, and fill the refrigerator again, but when Olanna "took long walks on campus, past the tennis courts and Freedom Square, she thought how quick leaving had been and how slow returning was." Their sense of place, of home is saturated by war.

Adichie's intention was not to write a lesson book about war in Africa; she is keenly aware of the dangers of didactic fiction. "The idea of using fiction to teach anybody anything fills me with absolute horror. Because I don't think it's the job of fiction. So, do I want to teach people about Africa with my work? No. Would I be happy if somebody who didn't know very much about an African country read my work and then started to think, 'There are human beings just like me in Nigeria and in Ghana and in Kenya and in South Africa?' That would make me very happy.

"One of the things that I hoped would happen when this book came out in Nigeria was that it would get us talking about this war that we just don't talk about, for a number of reasons," Adichie said. "I think that all countries are uncomfortable with the parts of their history that they find embarrassing or less than perfect. So I think that for a Nigerian, the war is something that still makes us uncomfortable. It's quite recent, when you think about it. The war ended in 1970. The major actors of the war are still prominent in Nigerian politics and I think a lot of the resentments are still festering. So we don't talk about it. Also, people are still divided about it. So the people from eastern Nigeria, which used to be Biafra, which is the section of Nigeria that seceded, experienced it in a way that was markedly different from the people who were in the other part of Nigeria. And so their memories are so different. . . . The history books just haven't figured out how to deal with it. We haven't learned much. For me, it's very gratifying to get an e-mail or to hear a story about [people] who, after reading *Half of a Yellow Sun,* go back and say to their parents, 'You have to tell me what happened to you, you have to tell me your own story.' "

IMAGINATION'S WELLSPRING

It was my tenth birthday; the package at the bottom of the stack of presents looked like it just contained a couple of books, so I opened it last. When I ripped apart the wrapping, I found that it wasn't books at all, but a little tape recorder, with three-inch reels and a separate microphone. The recorder itself was a creamy white plastic, the reels and microphone a deep red.

For the next few weeks I was rarely without my new toy. I quickly learned how to thread the thin brown tape through the machine and how to wield the mic. I convinced my friends to let me record them as we put on plays or pretended to be in the circus; chronicled my little brother's narratives as he played with his toys; captured my parents' heated discussions as they debated politics at the dinner table. I wanted to record everything.

Recently, my mother told me that she could see that I loved making things up, and had thought I might enjoy capturing all the make-believe on tape, but neither she nor my father had any ambitions that I'd become a radio producer when they chose that tape recorder for me as a birthday present all those years ago. She knew that, like most children, I also loved listening to stories; when my brother and I were small we devoured our father's made-up tales of three naughty chipmunks he named Oogie, Boogie, and Cahoogie. They lived behind our house and ended every sentence with an exclamation of "Oo oo ee ee ah ah ah!" And I spent hours dreaming over the beautiful illustrations in *D'Aulaires' Book of Greek Myths* while my mother read aloud about the passions and foolishness of gods and mortals.

These stories were captivating, but there was something more: there was the power of my parents' voices. They brought the characters to life, and connected me with the huge world outside our house. Perhaps most of all, they wove around me a powerful web of family intimacy.

My hunger for stories and love of voices (and perhaps that early facility with audiotape) led me to work in radio, a medium that captures the experience of listening. Radio is all about voices and stories: the urgent news of the day, the timeless yarns spun by master storytellers, the conversations that reveal lives and worlds. I find great pleasure in crafting radio programs, in part because I know that when my *Studio 360* team and I do our jobs right, our audience will feel the same eagerness and connection that I felt listening to my mother and father all those years ago.

Many artists have shared with *Studio 360* their own stories of people, events, and passions from their childhood that continue to influence their creativity. Filmmaker Mira Nair talks of her encounter with ancient myths and legends played out on a bare soccer field in a remote town in India; photographer David Plowden describes taking his first picture with a camera he was given as a Christmas present; and others, such as Richard Ford and Bill Viola, speak about the imprint of a childhood trauma that continues to surface in their art over a lifetime.

Each artist relates a particular childhood experience that

sparked a desire to create. Each was open to the strange and unexpected attractor of creativity—the sand in the oyster that gathers something powerful and beautiful around it. Each paid attention to it. The essential gift is to be open to wonder, and to let that wonder spark your creativity.

Filmmaker Mira Nair

Dreaming of a Wider World

"I used to love to listen to the radio in the rain."

There was just one movie theater in the town of Bhubaneswar in the province of Orissa, India, where Mira Nair grew up. "It perpetually and perennially showed *Dr. Zhivago*," she said. "It was weird, because sometimes the sound would go off or the power in the theater would go off and it was swelteringly hot and you would see just the image of snow and icicles on Omar Sharif's beard. And the manager would come on and say, 'I know it is hot here, but it's very cold in Siberia!' Anyway, it was a strange *Dr. Zhivago* each time. But really, I never knew of movies as an art form or something to be taken seriously."

Nair's father was a civil servant; her mother devoted herself to her children and to social causes. As a child, Nair spent hours on the streets of Bhubaneswar with her two older brothers. "We played cricket all the time, we grew up in rubble," Nair said. The surrounding buildings, however, were impressive. "In this little town there were two thousand temples," some more than a thousand years old.

The streets held other cultural treasures. "The major thing that made me a filmmaker was the traveling folk theater that would come through town." The players would act out epic mythological stories from the Mahabharata and the Ramayana. "I would go and see these great battles of good and evil [performed] by two people in a school field with no props but with a lot of passion, and hashish, as well. Everyone was stoned out of their heads, and it was amazing. After seeing that Jatra—the folk theater—I knew I wanted to perform."

At thirteen, Nair left Bhubaneswar to attend high school, where she had a part in every school play. "I went to an Irish Catholic boarding school in Simla, an all-girls school. I was always cast as the man because I had this deeper voice and I was tall. So I played Daddy Long Legs, and *HMS Pinafore*, and all these crazy musicals." She spent time in Calcutta, performing in political street theater, and at college in Delhi she began working with the Theater Action Group "run by Barry John, who was a disciple of Peter Brook. I joined his company and performed everything from antiwar theater to *Equus* to all kinds of things."

Nair was accepted at Harvard for her sophomore year of college. She wanted to pursue acting, but when she arrived in Cambridge, "onstage it was *Oklahoma!* and hoop skirts and musicals and things that I had nothing to do with." During school breaks she traveled to New York to observe her idols Liz Swados and Andrei Serban at the avant-garde La MaMa Theatre. When she returned to Harvard, Nair began taking photographs, and those led her to study documentary film. "I got hooked into cinema verité because it was a way of working visually, working with people, capturing something extraordinary in ordinary life," Nair told an audience when she received an award at her alma mater. "So I felt blessed that at the age of twenty I had found my place in life."

After graduating from Harvard, Nair moved to New York. She took a job as a waitress in an Indian restaurant to raise enough money to buy film stock. Over the next seven years, she made a series of well-received documentaries. The first portrayed an Indian immigrant's life in New York, where he worked at a newsstand at Nair's local subway stop, as well as his return to India to see his wife and new son. Next, Nair filmed *India Cabaret*, "inspired by the question I had in my mind about what divides good women from improper women in our society. . . . I chose to make a film about strippers in a nightclub in Bombay, and to see the world from their point of view. This was, of course, terrible for my parents." Nair loved the process of getting to know her subjects, filming their experiences, and creating a compelling narrative in the editing room. Her work met with acclaim around the world. But finding audiences for documentaries at that time was even harder than the grueling work of raising money to make the films. So Nair decided to make a fiction film about street children in Bombay.

She approached that project the way she had developed documentaries, by befriending children who lived in the city's streets. Weaving together the stories they told her with ideas she had been dreaming about, Nair and her college friend and film collaborator Sooni Taraporevala created a screenplay. They held an acting workshop for the children, many of whom became the stars of her debut feature film, *Salaam Bombay*. In this brilliant and heartbreaking story, Nair explores a childhood very different from her own. The film follows ten-year-old Krishna, who has to learn how to survive in the city alone. Krishna sleeps on rags with a group of street kids in an alley, and gets a job delivering glasses of tea with milk to the drug dealers and prostitutes in his neighborhood.

The movie is a bleak portrayal of the desperate conditions these children inhabit, yet there are moments of mischief and also delight that offer respite from the desolation of street life, suggesting, like the street performers of Nair's childhood, the power of art. In one scene, during a monsoon rainstorm, Krishna takes two cups of tea to a prostitute and her small daughter. As the woman dries Krishna's skinny shoulders with a towel, the radio plays music and the little girl says, "It's my song!" and begins to dance. Krishna dances with her, joyfully, with abandon, the music drawing together woman, girl, and boy, a moment when Krishna can just be a child.

Nair said that an experience in her own childhood inspired her to put a key detail in this scene. "They're listening to the radio in the rain, very evocative and very personal. I used to love to listen to the radio in the rain. I put a request show [into the scene] that recalled my childhood, and I actually put in the request the names of all my friends."

The detail of the call-in radio show evoked powerful responses from people who grew up in India listening, like Nair, to All-India Radio, as well as from many who had never visited the country. Nair believes that the more specific a story is, the more universal the resonance. "When you make a thing personal, it somehow speaks to people and the truth of it is felt."

Salaam Bombay was the first Indian film ever to win the *Camera D'or* at Cannes, in 1988. "It was a fairy tale, what happened to that film." Afterward, Nair began playing with another facet of her own youthful experiences to develop her next film. "I always had a second idea with me, which became *Mississippi Masala*, which is about what I call the hierarchy of color, and being a brown person between black and white. It was inspired by my years at Harvard University, when I came from India for the first time and easily moved between these two communities. Yet I understood there were invisible lines, as well, and wanted to make a tale with that. So with all the offers after *Salaam*, and all the agents and Hollywood and so on, I still kept to my compass and made *Mississippi Masala*, which was easier to get made because of the success of the first film, but was still quite a battle."

Nair cast the young Denzel Washington as the male lead, who falls in love with a young woman whose Indian family has settled in Mississippi after being expelled from Uganda. Nair said that at the time, "To have Denzel Washington in your film was not enough. A studio head said to me, 'Can't you make room for a white protagonist? Having an African American with an Asian girl is not, you know, I mean, it's all right, thank you, but . . .' And I said, 'I'll promise you one thing—all the waiters in this film will be white.'" And he laughed and I was shown the door and that was that."

Nair has since made a wide range of films. Some, like *Monsoon Wedding* and *The Namesake*, feature Indian families often straddling two worlds. In others, she looks at a different sort of outsider, such as Becky Sharpe in the Victorian English satire *Vanity Fair*, or Amelia Earhart, who won fame and success in what had been a man's world. Nair negotiates the territory between independent films and Hollywood pictures with finesse and humor, while also teaching at Columbia University in New York, where she lives. She makes it a point to return to India every year, and spends a few months each summer with her husband in his homeland of Uganda, where they hope to create an arts and cultural center in Kampala. Nair has begun with a workshop called Maisha, which has a yearly program for filmmakers from East Africa, where there is a large Indian diaspora. Every page of the Maisha website bears Nair's mantra: "If we don't tell our own stories, no one else will tell them."

Photographer David Plowden

Gathering Grist for the Mill

"When I started to photograph,
I did nothing but photograph locomotives,
because I knew they were disappearing.
That set in motion, I think, my whole career."

Intimate details of small towns, rural landscapes, and forgotten industry are the abiding subjects of David Plowden's evocative photographs. He has captured abandoned grain elevators in Arena, North Dakota; the fading, stately beauty of the Mid-Hudson Bridge in Poughkeepsie, New York; and the last brilliant sparks of America's industrial might in a steel mill in East Chicago, Indiana. Spend a few minutes studying the images in one of Plowden's many books of photographs about trains, bridges, barns, steamboats, steel mills, and small towns, or on his website's extensive archive, and you're immediately transported to a disappearing America, where black steam trains carve muscular lines in the Minnesota snow; tugboats undock steamships that resemble modernist sculptures; and copper mines cut brutal gashes in the Arizona landscape. Pulitzer Prize–winning author

David McCullough, who has written the introduction to many of Plowden's books, once said that Plowden's photographs "confer a kind of immortality on certain aspects of American civilization before they vanish."

Plowden traces his fascination with this subject back to childhood afternoons he spent gazing out the window of his apartment overlooking the East River in New York City. "In those days the river was just teeming with activity," Plowden told Kurt as they sat in Carl Schurz Park on Manhattan's Upper East Side, looking out at the same vista that Plowden loved as a child. "Boats up and down, and they had passenger steamers and they had tugboats and they had ferries and they had barges as well as oceangoing ships." Plowden's family lived at 25 East End Avenue, and when he was about five years old, the traffic on the river became even more spectacular. "For the 1939 World's Fair, the Hudson River Day Line and all of the passenger steamers ran boats from their piers to Flushing Meadows, where the fair was. There was a parade of these boats that went up and down all day long, and I was a little kid sitting at the window, mesmerized by all of this."

Boats were just one object of Plowden's childhood fascination. "At the north end were the great bridges, the Hell Gate and the Triborough Bridge, and to the south was that great pile of steel, the Queensboro Bridge. It was really quite a marvelous thing for a kid to look out on! I think it had a tremendous influence on the things that I've loved to photograph. I did a book on bridges and I know that it all started right here looking out at these wonderful structures. As a kid, you looked at them and you understood what made them stand up. I'm not an engineer, I'm hopeless, but looking at a bridge, you understand geometry. And the love of the steamboats and the tugboats started, looking down on the river and watching these things going back and forth. I did a children's book called *Tugboat*—what else would I do growing up on the river?"

As a young boy Plowden haunted the New York docks and befriended the ship captains. "I found that all of these men that I met on the tugboats sort of became surrogate grandfathers and fathers to me . . . I could tell by the whistles who was in the pilot house because they all had their own distinctive way of blowing the whistles, just as they did in the railroad. I was mesmerized. I still am."

In the summertime Plowden's family would leave the city for Vermont, where they had a farm. This led to another abiding fascination for the photographer: trains. "I spent an awful lot of time riding trains back and forth on the Central Vermont line; I knew all the engineers and I spent hours at the depot down in Putney. That was another life. When I was older, I got to know the railroad in New York, and I'd ride the Lehigh Valley and the ferries. I was a wharf rat when I was in New York, and I was a hobo when I was in Vermont."

When he was eleven years old, he was given a camera. Immediately, he decided to take a picture of the 4:20 train pulling into the Putney station. "The first time I went to photograph it, I got buck fever, and I handed the camera to my mother. I said, 'Here, you take it.' I started to shake. The next time I went down I was steadier, and I managed to take a picture. I still have it."

Perhaps Plowden hesitated because he was in awe of the power of these machines that breathed steam, and perhaps because, even then, he understood the enormity of the subject he was taking on. Yet in his youth Plowden didn't entertain dreams about becoming a photographer; all he wanted to do was ride the rails. His family had other plans; Plowden was sent to prep school—which he hated so much that he ran away—and later attended Yale. "I was going to be a railroad man, so I studied economics so I could go into the business of railroading. I very soon discovered that it was not the business end of railroading that interested me, it was the romantic end."

One summer during college, Plowden said, "I worked on the railroad pounding spikes, on the Delaware and Hudson on an extra gang outside of Oneonta, New York. And when I had a pocket full of money, what did I do? I went to Canada to chase steam engines." Just after graduation from Yale, he took a job on the railroad in Minnesota, not in the office but out on the line. "I rode all over the place, to the despair of my uncles and aunts and my mother's friends, who said, 'What's he going to amount to? He just rides trains.' And she said, 'I don't know what he's doing, but he does. Leave him alone; he's gathering grist for the mill.' She was my champion."

Whenever he could, he took pictures. "When I started to photograph, I did nothing but photograph locomotives, because I knew they were disappearing. That set in motion, I think, my whole career. I always felt that locomotives were closer to human quality than any other piece of machinery. And, you know, there was steam, there was hot grease, and you always felt they were alive.

"The people who worked on them felt the same way. They were all individuals—every single engine had its own personality. You'd talk to one engineer and he'd say, 'Oh, this one is a real liver splinter, it's very uncomfortable. This one is as smooth as cream.' Every one was a one-off creation." These huge machines fill many of Plowden's photographs; he captures the forward motion implicit in the wheels of Reading Railroad's Locomotive #2124 at rest in Shamokin, Pennsylvania, or finds Canadian Pacific's #5145 in the roundhouse in Montreal, surrounded by clouds of steam, its single light like a Cyclops's eye in the center of its imposing round face. "When the engine was in its roundhouse it was sort of like going into a stable, and being next to the horses. It's one thing to get them in the field, going by, and it's very dramatic. But go into the roundhouse and there's this intimacy. To me this is a portrait of a locomotive. I was walking around and around—it's a circle—and here's this great big wonderful face steaming away. How could you resist it? I was looking up at it just enough so that it would have that impressive overpowering look."

Plowden's images from the late 1950s chronicle the power and personalities of the engines and the sturdy sadness of the men who cared for them. A few years later, he depicts the loneliness of the empty depots and barren station platforms as this way of life died.

In those early years, riding the rails after college, Plowden was also smitten by the new landscapes he discovered. "I went one day to the end of a branch line in North Dakota. And I got out, and I walked beyond the tracks into the fields, and all of a sudden I realized for the first time that the world was spinning, the earth was spinning, in one of the flattest places in the world. There was nothing to hold on to. And I suddenly almost felt dizzy, and it was an extraordinary sensation. I worked for the railroad, and when I came back from the railroad, I said to myself, 'I have to go out there again.' I fell in love with grain elevators, which of course are the leitmotif of the prairie, of the heartland." These places are at the heart of Plowden's later decades of work, documented with a clear yet sympathetic eye in breathtaking photographs that employ only natural light.

Plowden has also pursued his fascination with the world of work and human creation in lyrical studies of bridges, of the devastation of mining, and of the towns, farms, and people of rural America. "I have been beset," Plowden has said, "with a sense of urgency to record those parts of our heritage which seem to be receding as quickly as the view from the rear of a speeding train. I fear that we are eradicating the evidence of our past accomplishments so quickly that in time we may well lose the sense of who we are."

Plowden's photographs are masterpieces that allow us to hold on to a sense of who we once were. His hand has been steady ever since his first aborted attempt to take a picture, yet you can still see the passion of that little boy in the pictures taken by the man.

Novelist Richard Ford

Weaving Childhood Stories

"Lots of things I choose to write about, inevitably they're something I've known about for years but that for some reason or another never migrated into my imagination as something I could use."

In 1996, Richard Ford won both the Pulitzer Prize and the Pen/Faulkner Award for *Independence Day*, which takes place on the Fourth of July. It was his second novel about the life of Frank Bascombe (the first, *The Sportswriter*, is set during an Easter holiday). Ford's Bascombe was a novelist who abandoned fiction for sportswriting when he was in his thirties, and then abandoned sportswriting to become a real estate salesman. Over the course of the first two novels, he wrestles with the death of his young son and the end of his marriage.

In 2006, Ford published the final part of his Bascombe trilogy, a novel he called *The Lay of the Land*. Set just before Thanksgiving in 2000, this novel finds Bascombe in what he calls "the Permanent Period" of life in one's fifties, driving around central New Jersey, selling houses and ruminating on the botched presidential election, real estate values at the Jersey Shore, and the vicissitudes of back roads and backwaters in the central part of the state. Bascombe also mulls over what has become of the essential characters in his life, including his grown son and daughter, the son who died in childhood, his Tibetan real estate partner, and his estranged second wife.

In the middle of the novel, Bascombe indulges an ex-girlfriend's ancient and ornery father in the old man's favorite pastime: viewing the demolition of hotels in Asbury Park. While he watches the structure of a hotel implode, Bascombe is caught up in an ancient memory of an afternoon when his mother unknowingly ran over their new kitten, Mittens. Just after *The Lay of the Land* was published, Richard Ford explained that this image comes directly out of his own childhood, when, in 1950, Ford's mother pulled out of their driveway in Jackson, Mississippi, and accidentally killed the family cat. Richard, her six-year-old son, was the only witness.

"She didn't know she ran over the cat and just backed out into the street and drove away and I was left alone at the house with the cat in its death throes," Ford said. "Obviously traumatic, okay, but a lot of traumatic things don't make their way into your book. . . . I picked the little cat up and I ran next door where there was this old Civil War veteran—you can imagine that, a Civil War veteran, at least he said he was—sitting on the front porch. A very, very, very old, turkey-necked man. I ran up to him and I said, 'My cat, my cat's been run over!' And he was of course as addlepated as could be and he said, 'Oh, I think you've got a Persian there, but it looks like something's wrong with it.' I've known that story all my life, but I haven't ever thought that it would ever find its way into something I wrote. I would sometimes tell it to people—I've told it to people for decades. Suddenly a moment comes in the middle of this book where it seemed to fit in perfectly, and there it was.

"Lots of things I choose to write about, inevitably they're something I've known about for years but that for some reason or another never migrated into my imagination as something I could use. You could also say that the pantry was getting bare, and I was looking back behind the first row of cans. For me, even if they're not cataclysmic events—9/11 for instance—events around me have to settle into the ground and then percolate back up through my feet."

Ford grew up in Mississippi in the 1950s, and spent a lot of time in Little Rock, Arkansas, where his grandparents owned a hotel. "It was a great way to grow up. Because it was a sort of anti-Eloise experience. Big old blowsy drummer's hotel that also housed the state legislature when the legislature was in session in Arkansas. Governors came there all the time. My first job was as a room-service waiter. You get to see at the age of sixteen what adults do when they don't want to be seen doing it. And I saw a lot of that—knock knock knock, the door opens, here I am, there they are, you see some interesting tableaus."

Ford has written essays about life in that hotel but doesn't think he'll ever write fiction set in Arkansas. "I am going to write a novel which is set in Saskatchewan about a man who runs a hotel out on the plains there, so maybe some little flecks of that experience of younger life could find their way into another book." Since his early work, Ford has generally left the South alone. "When I was thinking about writing the first book, which was set in the South, thirty years ago, I was thinking that I could write a book which would break the mold, which would be set in the South but wouldn't be thought of as a southern novel. And the first thing anybody ever said and the last thing anybody ever said about it was 'another southern novel.' And I thought, I don't want to do that, I don't want to write just for southerners, I don't want to write just on southern subjects, or set books in the South. I really want to write books that break some kind of mold."

As a young boy, Ford didn't dream of becoming a novelist—in fact, he wasn't much of a reader. "I was dyslexic and I read hardly at all. I learned seriously to read in my early twenties. Of course, I *first* learned to read along with the other second-graders, but I was slow. I went all the way through school not really reading more than the minimum, and still to this day can't read silently much faster than I can read aloud. But there were a lot of benefits to being dyslexic for me, because when I finally did reconcile myself to how slow I was going to have to do it, then I think I came into an appreciation of all those qualities of language and of sentences that are not just the cognitive aspects. The syncopations, the sounds of words, what words look like, where paragraphs break, where lines break, all the poetical aspects of language. I wasn't so badly dyslexic that I was disabled from reading; I just had to do it really slowly. And as I did—lingering on those sentences, as I had to linger—I fell heir to language's other qualities which I think has helped me write sentences."

The first book that really captured his heart was Faulkner's *Absalom, Absalom!*, which Ford read when he was nineteen. "It's probably the first book I ever fully read. It struck me that I had been reading up till then trying to absorb facts, language as communication. Richard Hugo has a wonderful little phrase in his book *The Triggering Town*, which is about poems. And in the essay he says, 'when language is just a form of communication, it is dying.' And so up to then, in my twentieth year, I had just been reading as you read the newspaper. But poring through those long Faulknerian sentences, I had this kind of inkling, 'This has to be about something else, because it's taking me so long to get through it. I'm wading through these phrases: Is there something here that I'm not seeing that's causing me to have to go through this so ponderously and to be pleasurably aware of so much more?' And I thought, 'Yeah—the mode of address, the form of the sentences, the style of the sentences, how words sound, those are part of what literature is. They express the part that newspaperish "communication" can't.' And when I became attuned to it I realized that it was there to give me pleasure and to cause me to know the otherwise inexpressible in life."

Ford said that when he was writing *The Lay of the Land* he read the entire novel out loud to his wife, Kristina. "She helps me immeasurably . . . We started [in] January and steady on read it straight through, she with a manuscript, me with a manuscript, sitting across the room from each other in our rented apartment in New Orleans, right to the end. It was quite a gesture of affection on her part to let me do that; because she would stop me and I would stop myself and we would bicker about the lines. It was always to the work's benefit. And she'd say, 'Read that again, is that the word you want to use?' She's always helped me, we've read all these books aloud. I feel like if I don't read things aloud, I don't really fully authorize them, I have to hear everything, hear what every sentence sounds like. I write so somebody will read what I write. And she's my audience. And a good one." Ford's sentences often contain internal melodies that meander and repeat, and whether you read his words on a page or listen to Ford speak or read aloud in his lovely southern cadences, it's clear how much the music of language means to him.

Video artist Bill Viola

Returning to a Moment

"It was a place I try to get back to my whole life."

In Bill Viola's beautiful and mysterious videos, two things are a constant—the intense slow motion, and the artist's play with elemental forces. Images in his emotionally resonant and critically acclaimed work often include bodies engulfed in flames, or people falling or rising through water and toward light.

One can look at all of the water in Viola's work as a metaphor, as a sensuous material that the artist has used in many different ways; it's also something that has captivated his imagination since a summer afternoon in 1958 at Trout Lake in New York State, when he was about six years old. The artist described "jumping off the raft on the lake on summer vacation, and forgetting to hold on to my inner tube and sinking right to the bottom like a stone. And after some time—I don't know how long—my uncle dove in and grabbed me and saved me, and I was sputtering and crying. But I had glimpsed another world that stayed with me for my whole life. I can still see it now and feel it. It was the most peaceful experience I ever had. A place I wanted to stay in. I had no fear, even though I was drowning. And it was a place I try to get back to my whole life; I guess I do that through all this water imagery."

The imagery that Viola has created is both beautiful and scary. Take, for example, his installation for the Venice Biennale in 2007, called *Ocean Without a Shore*. Above the three imposing altars in the Church of San Gallo, where one might expect a Titian Annunciation painting, Viola placed three large plasma screens. Even when viewed on a small computer monitor, the images Viola projected there are captivating; at first black-and-white, with hazy outlines of people, one for each altar, filmed with an old bank surveillance camera. There is a sense that they are walking toward the observer slowly, slowly, slowly. Suddenly, one figure, a woman, raises her hands, pushing through a previously unseen wall of water as she walks, her gray dress becoming bright red, her blurry features sharpening, wet, as she passes from gray into color, through a scrim of cascading water. "They walk through a wall of pouring water which is so thin and transparent that you can see through it, you don't even see it," Viola explained. "All of this water flares up off their bodies and hits the light, and then at the same time they transition from this thirty-year-old video technology, to the latest, best highest-quality, high-definition camera where you can see every little drop on their body, you can see their soaked clothes and their dripping hair; they literally become incarnate. Which is the idea of the piece. They stand before the camera; it's usually a very emotional experience passing through that water—the whole crew and the actors, everybody was very, very deeply moved each time someone would do it. Then they stand here and look at us, and then they realize they have to go back, because their time here is temporary. And they turn their backs, literally, on the world, and walk back through the water, and become a grainy black-and-white apparition and eventually just disappear into the darkness."

The effect is mesmerizing, and not a little spooky. Viola said that this piece is "about the presence of the dead in our lives." When he was asked to create something in this beautiful church, at first he had no idea what he was going to do. "So I began to really focus on these altars and realized that, as per the original development of the origins of Christianity, these altars actually are a place where the dead kind of reside and connect with those of us, the living, who are here on earth. . . . This is a piece about humanity, and it's about the fragility of life, like the borderline between life and death [which] is actually not a hard wall, it's not to be opened with a lock and key, it's actually very fragile, very tenuous."

In much of his work, Viola examines this thin membrane between life and death, submersion and emergence. His images are haunting and brilliantly gorgeous, and perhaps trace their origins back to that summer afternoon long ago. "I guess my experience in the water was beautiful until I got pulled out of it and realized what had happened," Viola said. "Then I started crying. Two aspects of human existence in our lives that attract me are violence and beauty, and very often they are tied together. And I don't just mean violence like the horrible stuff you see on the news. For example, when my mother passed away, I was with her at her bedside, holding her hand. And I looked at her face and it was one of the most beautiful, profound things I'd seen in my life. My own mother who had just left her body moments before and all of a sudden there was this face, staring, and it was like it was angelic almost. Even though most people would describe it as being horrible or terrifying, for me it was deeply, deeply profound."

In 1991, Viola created *The Passing*, weaving together images of his mother as she lay dying, her breathing labored as we view her on her hospital bed, with video of Viola's son being born. Viola himself appears in the film, submerged in water. *The Passing* is another of Viola's lyrical evocations of the fluidity of life and death. "Even though I use this media, I think the visual aspect of it is secondary—that to me is not the main part of it. I've been striving to turn my camera on the kinds of things you can't see."

Chapter 6

MOTHERS AND FATHERS

If you had opened the freezer in my family's kitchen when I was growing up, you were just as likely to discover a plastic bag filled with a thick, typewritten manuscript next to the frozen broccoli as you were to find a carton of ice cream. My mother, Janet Handler Burstein, went back to school to get her M.A. and Ph.D. in English literature when I was in grade school, and she kept drafts of her dissertation in the fridge. This was long before personal computers, and someone had once told her that the freezer was the safest place to store her work if the house were ever to catch fire.

When I think back, I see Mom sitting at the dining room table in her writing uniform of blue jeans and one of my father's faded old shirts, surrounded by towers of red-bound library books, typing a million words a minute on her IBM Selectric. A few years later, she's at her desk, grading papers or preparing lectures in the light-filled office that my parents built onto the house when she became a professor.

Over the years, my mother and I have had a wonderful, complicated, occasionally fraught relationship filled with both tenderness and fury. We've shouted at each other on street corners; there have been times when one of us refused to answer the other's phone calls. We've also held each other up through various family illnesses and crises, and have spent hours together talking about concerts and movies and books and art.

And yet we have made perhaps our strongest connection through work. She no longer teaches full time, but Mom continues to write articles and criticism, and will sometimes ask me to read a draft of a piece before she sends it off, which I do with pleasure. I often turn to her when I'm struggling through a thorny problem with a radio story, and she's suggested a number of ideas and guests for *Studio 360* over the years. Somehow, while wrestling with words on a page or voices on an audiotape, we connect as our best selves. This wasn't always the case. There were years when I needed to find my own way and it was too threatening to consult her. But it's a gift that we have discovered a new connection with each other at this point in our lives, and she was my first reader for what you're reading now.

What is it about the dynamic tension between parents and children that can spark, or sometimes smother, creativity? In *Studio 360* artists have talked about a few of the powerful ways in which this central, complex relationship resonates in their creative lives. Some, like the Tony Award–winning actress Patti LuPone, tell tales of disappointment and misunderstanding. Writer David Milch reflects upon his father's influence on his ability to create complex, sometimes violent characters who are nonetheless worthy of love. The actor Kevin Bacon admits that he was "driven to become more famous" than his father, the brilliant and prickly Philadelphia urban planner Edmund Bacon. Rosanne Cash reveals an intimate moment of conflict and resolution with her iconic father, Johnny Cash. And saxophonist Joshua Redman describes growing up without his father, but with his father's music.

The stories in this chapter are filled with mixed emotions: love and frustration, irritation and affection. These artists articulate, through their work and the stories they tell, the profound connection that began when they were born, and which continues long after their parents are gone.

Actress Patti LuPone

Born for the Stage

"I knew at four where I was going, and just did it. [But] I had to drag my mother."

Filial tensions often take center stage in theater, from the tragic father-daughter relationships in *King Lear*, to the archetypal stage mother in the musical *Gypsy*, based on the autobiography of burlesque star Gypsy Rose Lee. Both the book and the musical explore the tumultuous relationship she had with her mother, Rose, who pulled her two daughters into the spotlight almost before they could walk. Yet both works also look behind the facile image of the pushy stage mother to glimpse the intricate dance between parent and child.

This is clear in the final scene of the musical. When her mother comes into her dressing room after a show, Gypsy, who is now famous, throws her out. Storming onto the empty stage, Rose sings her final showstopper, "Rose's Turn," about the fierce disappointments of having ungrateful children, and of never having lived her own dream.

The role of Rose is both a challenge and a gift for an actress, and, in 2008, Tony Award–winning Patti LuPone got her turn. She can picture exactly where she was when she first heard Ethel Merman, the original Mama Rose, sing "Rose's Turn": LuPone was four years old, listening to the original cast album LP in her living room on Long Island. "I remember growing up and listening to Ethel Merman and Mary Martin," LuPone said. "I was born to be on the stage. . . . I knew I had a big voice; having heard those albums, I knew that my voice was built for the Broadway stage. I knew at four where I was going, and just did it. [But] I had to drag my mother."

LuPone was a stage kid, and her mother played her part. "My mom and dad got divorced . . . and my mother threw everything into her kids," LuPone said. "That became her life. She just drove us around," from lesson to rehearsal to performance and back.

LuPone and her brother Robert both grew up to be stars, which seemed to come as a bit of a surprise to their mother. "She famously said to Bobby and me, 'When are you gonna stop flitting from job to job?'" LuPone said. "Like we could get a thirty-year musical or something! She was a great supporter but had no talent of her own. And didn't understand where we got ours, or what our drive was."

A child's urge to be acknowledged for who she really is runs deeper than many parents realize. LuPone says it informs her performance of Mama Rose, all these years after her mom shlepped her to rehearsals on Long Island. "My mom and I did not have a great relationship, because I don't think she understood me. In the discovery of this part for me, I conjure my mother." The strains in that relationship, she said, help her understand Rose, and allowed LuPone to put her heart into Rose's final lament, which, she said, is "a parent's scream."

Graphic artist Miriam Katin

Telling a Mother's Story

"It's sort of a daily unwanted, uninvited reality, a narrative going through my head."

It took some lubrication for Miriam Katin's mother to read through an early draft of Katin's first book, *We Are on Our Own.* "She was very concerned that someone will take offense, and so I sat her down with a nice glass of scotch," Katin said, "and I let her read. And when she was done, she turned around, she was crying, and she said, 'You have done a beautiful thing here.' I had the freedom to finish. I can't even tell you what a relief that was for me."

An accomplished artist who had illustrated children's books and worked as a background designer for animation at MTV and Disney, Katin published *We Are on Our Own* in 2006, creating pencil drawings, mostly in black and white, with occasional bursts of color, to portray the story of how her mother hid them both from the Germans during 1944 and 1945. Katin was a tiny girl at the time; she was born in Budapest in 1942. "What I remember really was mostly about food," she said. "A little dog eating my feces, and bombing, and that was about it. . . . The bulk of the story I didn't find out until I was thirty years old [and pregnant] with my second child. My mother sat there and told me this whole story, and I was shocked."

Once she knew the story, Katin couldn't let it go. "It's sort of a daily unwanted, uninvited reality, a narrative going through my head. But I kept saying, 'Well, another Holocaust story. Who needs it?'" The success of Art Spiegelman's *Maus*, about his parents' experiences during the Holocaust, spurred Katin to tell the story of her mother, whose chutzpah and determination are stunning as she relinquishes everything she's always valued in order to survive.

The first pages of *We Are on Our Own* show the comfortable, cosmopolitan life that Katin and her family enjoyed before the German invasion. It's a brilliant fall day, and two slender young women are drinking espresso at an outdoor café in Budapest. They are beautiful in their stylishly tailored suits with broad lapels and knee-length skirts, high heels, and hats set just so on their carefully coiffed heads. A tiny toddler and a dog play at the women's feet. Life is clearly lovely. But a few pages later, Katin's mother, whom she calls Esther in the book, receives a notice demanding that she and her daughter join the deportation of Jewish citizens.

Esther's friend gives her the number of a black marketeer so that she can procure false papers saying she's a peasant with an illegitimate daughter. She burns anything that might reveal their identity: photographs of her mother and father, letters from her husband, Hebrew prayer books. She instructs her maid to tell the neighbors that she drowned herself and her daughter in the river. Dressed now in a long, loose skirt, rough jacket, a scarf over her head, she holds the little girl's hand as they walk toward the soaring glass-roofed Budapest train station. "Good-bye, cosmopolitan lady," Esther says to herself. "I must act this new role, must not look scared. Look defiant."

In the station, mother and daughter walk a gantlet of soldiers watching people leave Budapest. The men taunt and tease, "Hey, cutie!" and Esther, a smile on her face, responds, "Kiss my ass, stupid." Katin said her mother told her, "'I don't know why I ever made that route with you. I was trying to behave like a peasant girl, and they were whistling and saying dirty words, and so I shouted back.'"

Throughout the book, Katin depicts the desperate choices her mother had to make in order to save herself and her small daughter. They seek refuge with an old couple who grow grapes, but are discovered by an SS officer searching for some good wine. "Such a beauty here in no place" he says, and returns with chocolates for the little girl, whose mother must trade sex for safety. Soon the Russians arrive, and rape the women they find in the farmhouse. With soldiers in pursuit, Esther runs out of the house into a snowstorm, child in one arm, suitcase in the other, struggling against the wind until she places the suitcase on the snow, and puts her small daughter inside.

It's a scene that is seared in the author's mind. "Strangely, until I finished the book I couldn't talk about it because I would choke up," Katin said. "That was when she had to run for our lives in the winter and she couldn't carry me so she put me in a suitcase and she was pulling me. . . . That image was with me so dramatically." We can see it: in spare, evocative black-and-white drawings, a small girl grasping the sides of the suitcase as her mother, black hair streaming behind her in the wind and snow, pulls her to safety in the woods.

Katin portrays the battles that rage around the countryside, as well as the buds on the trees as spring arrives and her mother tries to make their way back home. In a small country train station late at night, a Russian soldier shows her a picture of his own family, so Katin's mother won't fear him and will be able to sleep until the train comes. Another soldier helps Esther and her daughter board the train the next morning, and as he lifts them up into the compartment he says, "Remember Russian soldiers good men."

This undercurrent—that in each desperate situation, a decent stranger helps them—is very powerful. Katin believes that acknowledging this was central to her mother's ability to survive after the war, with all its horrors, was over. It is also a gift that her mother gave to her. "This is my mother's big mantra, that not everyone was an animal," Katin said. "There were people who were good. She kept saying, 'If you keep hating, it will destroy you from the inside out. For those people who were good, we must forgive.' My mother, by one way or another, kept me innocent of all of them, and allowed me to be a rebellious teenager. I never got that 'Oh, what I did for you.' She allowed me to be a little bitch sometimes. Even now I have some difficulty mother-daughter. She lets me have that freedom."

Miriam Katin's mother lives in New York City and still takes the subway to deliver home-cooked food to her grandchildren. Katin dedicated *We Are on Our Own* with an inscription to the woman "who taught me to laugh and to forgive." In *We Are on Our Own*, Katin captures a family trauma, within which she finds both strength and terror. It's almost a fairy tale: the brave mother and tiny precocious daughter facing evil, finding good people who help them, and surviving.

Writer David Milch

Loving a Complicated Father

"To be able to testify to the complexity of a personality at such a level that he or she is always demonstrated as worthy of love, no matter what behavior he or she is engaging in, is the privilege of my professional life."

It's not always easy to see the autobiographical layer in a fictional work, although that dimension is often there. There is no one named "Elmer Milch," for instance, in any of David Milch's television dramas, such as *NYPD Blue* or *Deadwood*. What you will find are flawed heroes, such as Andy Sipowicz, the alcoholic, racist police detective on *NYPD Blue*; and powerful protagonists, such as Al Swearengen, the brutal, foul-mouthed owner of the Gem Saloon, who runs the frontier town in *Deadwood* with a knife in one hand and a bottle in the other. But Swearengen also rushes to help when the preacher falls down in what looks like an epileptic fit, putting a wallet between his teeth to keep him from biting off his own tongue. What is extraordinary about characters like Swearengen and Sipowicz is that in addition to the antisocial aspects of their personalities, there is a redemptive humanity beneath the brutality. And because of this, in addition to being riveted by these fascinating characters, we develop compassion for them, too.

Milch has often said that these deeply complicated portrayals owe much to his father, who was a respected surgeon and teacher in Buffalo, New York. But Elmer Milch could act violently toward his children. He was a gambler and an addict, vices that would seduce his son as well. Milch described his father's uncles as being "in the rackets," and every summer, the elder Milch would take David and his brother and mother to Saratoga, to the racetrack, where all of his family would congregate. During the Kefauver hearings on organized crime in the 1950s, Milch says that twenty-five "wiseguys" lived in their house in Buffalo because his father was giving them all hernia operations so they wouldn't have to testify.

Milch's father sounded like a fictional character himself: he had mobsters as friends, he beat his kids, and yet people loved him. Milch responded: "And I did, too. . . . I loved him. To the extent that you're making a connection between Swearengen and Sipowicz and my dad, it is the great joy of my imaginative life to make that journey myself. Because there was no moment during the time when my dad was alive that I didn't love him. No matter what was going on. And to be able to testify to the complexity of a personality at such a level that he or she is always demonstrated as worthy of love, no matter what behavior he or she is engaging in, is the privilege of my professional life."

The characters and worlds that David Milch creates are deeply compelling, often disturbing, and capture a truth about the people and places he describes. When asked if he had anticipated, before the show began, that Al Swearengen would become the glorious center of the town of Deadwood, which is populated by foul-mouthed miners and prostitutes and murderers and people hoping to get rich just outside of the reach of the U.S. government, Milch replied, "I guess Milton figured that Lucifer would get the best lines. Simply because there's more scope to the departure from goodness, in someone who is fundamentally good, there's just more scope in that character. And so I kind of had that sense, the same way I felt that Sipowicz was going to be in *NYPD Blue* a more interesting character. It's harder, but much more gratifying, ultimately, to learn to love a character like that. And people like to go on journeys, and that's the best kind of journey to go on."

Actor and director Kevin Bacon

Competition and Loving Neglect

"He was definitely not the type of guy who would take you out on a Saturday afternoon and throw the ball around with you."

In the second film he directed, titled *Loverboy*, Kevin Bacon explores how the failures of parents can damage two generations. He plays a father who is so wrapped up in his relationship with his wife that he has no time for his daughter, who grows up to become consumed with the desire to have a child. But once she does, she finds she is tragically incapable of sharing her son with the world.

"Both as a director and an actor, I'm much more into gray than I am black and white," Bacon said just after the movie was released. He was fascinated with "this idea about how wrapped up a woman can be with a child, and how important a child can be to a parent, and how far does that go? If you're walking down the street, there's a strange thing when you have kids, and the first time they let your hand go, and they shake your hand off, and run down the street to spread their wings a little bit. It's very bittersweet. You kind of say to yourself, 'I guess this has to happen in life,' and it continues to happen in stages all along. But there is something that also makes you want to hold on to them. So the movie investigates a darker side of that personality."

This was the third film in three years in which Bacon explored the dark corners of childhood and parenting. In the 2004 movie *The Woodsman*, Bacon was able to find a deep humanity in a character most people would find abhorrent, playing a man who returns home after twelve years in prison for molesting young girls. And in the 2003 film *Mystic River*, he plays a cop who has to investigate the murder of a childhood friend's daughter. The main suspect is another friend, with whom they share a disturbing past, because when they were all eleven years old, this friend was abducted and molested.

"I haven't really gotten on the couch about that one, but my guess is that when you have kids, you do live with a certain amount of fear," Bacon said about his choice of roles. He has two children with his wife, actress Kyra Sedgwick, and the whole family appears in *Loverboy*. "You live with that gnawing thing of 'What if something bad were to happen?' And that would be the worst thing that you could possibly imagine. Sometimes you just have to, in your work, confront those fears in some way. I never set out with those three [films] to look for bad things that happen with kids; it's just sort of laid in that way. That's the only explanation I can come up with."

Bacon spent his own childhood in Philadelphia, where his father, the urban planner Edmund Bacon, dramatically transformed the landscape of the city during his more than twenty-five-year tenure as the executive director of the Philadelphia Planning Commission. "He was busy, and certainly very busy by the time I was born, because I was the youngest of six," Bacon said. "So, as my brother likes to describe our relationships with our parents, it was 'loving neglect.' He was definitely not the type of guy who would take you out on a Saturday afternoon and throw the ball around with you."

Nevertheless, Bacon said that when he and his siblings were growing up, his father and his mother, Ruth, were very supportive of anything creative. "There was never any kind of pressure on us to get good grades," he noted. "There was never any kind of pressure on us to make money. There was a lot of pressure to create something from nothing, take some Elmer's Glue and some matchsticks and build a little house, or put on a play or write a song or play an instrument, whatever it was. That was the directive in our house, and that was helpful for me as an actor."

A towering public figure in the Philadelphia cityscape, Edmund Bacon was both beloved and feared for the changes he wrought and for his outspoken views and often cantankerous personal style. "My father was a very, very big personality. He was a very eccentric man, and kind of brilliant in his own way. And charismatic, absolutely. And incredibly driven. Not at all a typical father, and my relationship with my son is hopefully a lot different from my relationship with my father."

Bacon, who starred in films like *Footloose*, *JFK*, and *Apollo 13*, is also renowned as the central character in the trivia game "Six Degrees of Kevin Bacon," developed after the concept that anyone in the world is connected to anyone else through six relationships—or less. He's had his share of fame, but in his hometown he often wasn't the one who was noticed. "Down in Philadelphia, we'd walk around, and right up until a few months before he died, I'd walk with [my father] down the street and I'd hear someone say, 'Mr. Bacon,' and I'd turn around, expecting to sign an autograph or take a picture, and they'd be screaming to my father. He was very well-known down there; he really was a star." Perhaps that is Edmund Bacon's most enduring legacy for his son; Bacon said that his father "was very inspirational to me, in that I think I was very driven to succeed and driven to become more famous than him." Outside of Philadelphia, Kevin Bacon has definitely succeeded.

Singer and songwriter Rosanne Cash

Healing Onstage

"I felt I just needed to get out of his shadow
and do this on my own."

The attention that constantly swirled around her father, Johnny Cash, was a compelling reason for Rosanne Cash *not* to become a musician. "Why would anyone have a career where they might get famous, the worst occupational hazard in the world?" she wondered. "For many years I thought this was the last thing I wanted to do."

Cash said that as a child she hated the demands attached to her father's fame. "That my dad was always away, that he was stretched to the maximum in terms of his stamina and energy and emotional health and his physical health, and he had constant intrusions on his privacy. He couldn't go to a movie without people bothering him. I started thinking that all these people who constantly came up to him were figures from my subconscious, or maybe his subconscious, and they were just specters sent to intrude."

But Cash also knew from the time she was a very little girl

that she wanted to be a writer. "When I started writing songs, I had this agonizing period where I thought, 'Well, am I just going to write songs for other people or do I want to do them for myself?'" Her reluctance wasn't because she thought her work wouldn't stand up to the music of her father, to whom she knew she would be compared. "I had a lot of hubris about being a songwriter. From the very first songs I wrote, I thought they were great. It took me a few years to realize how horrible they were. But about going onstage—that took a long time to accept. I just didn't want that much attention. By nature I'm private and really shy and I just thought, 'Why would I do this?' I struggled with myself for years about doing it."

At eighteen, she finally told her father what she wanted to do, and showed him some of her songs. "He was incredibly encouraging. He would never say, 'Oh, well, you need to work on that.' He was actually just a total dad in that situation. 'Oh, you're great, honey. That's wonderful, honey.' Then when he saw I was going into this, that this would be my career, he got a little nervous, saying, 'Well, you know, just take care of your children first.' That was really the only advice he ever gave me."

Her stepmother, June Carter Cash, "was key. I don't think I would have been a performer without her. She and her sisters were the ones who taught me to play guitar. And Maybelle, her mother, taught me the family songs. I still play like they do, with that little Carter scratch. She would say, 'Oh, you don't have much of a voice, but you're just gonna be great.' She was fantastic."

Rosanne Cash recorded her first album when she was twenty-four, and over the next ten years had both country and crossover hits. She won Grammys, was named top singles artist by *Billboard*; it was a time of great success. It was also a period during which she withdrew from her father, as she worked to establish herself on her own terms. "I felt I just needed to get out of his shadow and do this on my own."

There were also ghosts to wrestle. "I [went] through a period in my life where I was angry with him. All of the childhood resentment had surfaced at that point and I had stuff to talk with him about, and he asked if I would sing with him at Carnegie Hall. I was angry and said, 'No, I've got a headache. I don't want to.' He just nodded and said, 'Okay.' The day of the show I came into his hotel room and he asked me again. And again I said no. I was still angry. And he said, 'Okay.' He just accepted it. He got up and walked away, and there was just something about the look of his back that I had seen onstage a million times, that back framed in spotlight, and it just broke my heart. I called after him and I said, 'Dad, I'll do it.'

"We sang 'I Still Miss Someone' that night at Carnegie Hall. It was a transcendent moment; it was like everything was washed clean. And it made me realize that the stage is where my dad worked out all of his deep problems, and where he got healed. And there was a space to contain me that night. It was beautiful, it was a really sweet moment."

Saxophonist and bandleader
Joshua Redman

Reconciling Influence and Distance

"I didn't grow up with my father,
but I grew up with his music."

As he was preparing to record his eleventh album, Joshua Redman asked his father, saxophonist Dewey Redman, to play with him. "This was the first time I got up the courage to ask him. I didn't know if he would agree."

His father said yes, and the album became *Back East.* On one tune on the record, called "India," father and son square off, one on the left, one on the right. They engage in a dialogue with a joyous yet competitive call-and-response, each trying to outdo the other, finally blending into a single vibrant note, with Dewey Redman's beautiful vibrato and Josh's deep resonance, before the rest of the band joins in. In this music, father and son feed off each other, call forth the best in each other, show off and then relax into a lovely unison.

Joshua Redman grew up as Joshua Shedroff in Berkeley, California. His mother, Renee Shedroff, was a dancer who became a librarian so she could support herself and her son. By the time Joshua was born, his father, Dewey Redman, had moved to New York to play with other jazz greats such as Ornette Coleman, Charlie Haden, and Keith Jarrett, and to eventually lead his own band. "I didn't grow up with my father, but I grew up with his music," Redman said. "I had all his records and would hear him when he'd come to town. But I spent very little time with him personally. I never felt on a conscious level that I was following in my father's footsteps. I never thought, 'Oh, I'm playing the saxophone so I can be like Pops, so I can carry on the family legacy. Of course, a lot of other people seemed to think that, but I just thought, 'Ah, I like the sound of the saxophone.' Now, who knows what was working on that subconscious level."

As a child, Redman also listened to a lot of Asian and African music. "My mother was particularly interested in Indonesian dance and Indian dance and African dance. So I grew up with these sounds." Those influences can be heard in some of the tunes on *Back East*, whose title is a clever play on the East that is New York, where it was recorded (Redman still lives in Berkeley), and the East that is beyond the oceans.

Redman played sax in his high school jazz band, and then went to Harvard, where he graduated Phi Beta Kappa. He was poised to attend law school in 1991 when he decided to take a year off to play music in New York City. That fall he won the prestigious Thelonious Monk Jazz Competition, and started to get attention from record labels—and from his father. "We had a complex relationship, and we went through periods where we were closer than others," Redman said. "When I first moved to New York, right out of college, that's when I really started to get to know him. I was playing in his band and touring with him. That was probably our closest period.

"And every night we would play something very free-form, but we'd also play bebop, we'd play a real straight-ahead tune, we'd play a ballad, we'd play a down-home Texas blues. So he liked to play a little bit of everything. His sound was so strong and so deep, his personality would carry through in whatever style he played. You always knew it was him.

"After I first moved to New York, a lot of people tended to comment on how little they thought I sounded like my father, and I always thought that was strange because he has been one of my greatest influences. I've been listening to his music since the day I was born. And I've always felt there was a lot of his sound and his playing in what I've done. But it's true that on the surface, we had very different styles." It was while living in New York that Joshua legally changed his name from Shedroff to Redman.

In 1993, Joshua Redman's first album as a bandleader was nominated for a Grammy. "After I started to do more of my own things as a leader, I think we drifted apart for a little bit," he said. Joshua's success was bittersweet for Dewey Redman, who in 2003 said that his son was "a very talented young man. Very well spoken and plays his ass off. I'm better, but he's gotten all the things I never got. I'm not jealous, just a little bit envious. But I'm very proud of him."

For Joshua Redman, his father is "one of the greatest saxophonists, in my opinion, of all time, but certainly of his generation. [He had] universal love and respect within the jazz community. But he certainly struggled and didn't have a lot of the opportunities that many of his peers had, and didn't have the opportunities that I had. I burst on the scene, all of a sudden had a record contract and was able to lead bands and make a good living. And I think he was tremendously proud, but it wasn't fair, I didn't think it was fair, there was a mixture of a lot of emotions on both of our parts."

Dewey Redman joined his son in a New York studio in the spring of 2006 to make Joshua's album *Back East*. After they had finished recording the tunes Joshua had written, Dewey said he wanted to record something else, by himself, with only bass and drums. "I walked around the corner and got an espresso," Joshua recalled in notes on his website. "I came back maybe eight minutes later, and Dad was already packing up. 'How'd it go?' I asked. 'Fine, one take,' he said. So I thought, 'Great. Let's move on.' We were kind of running behind in the session, so I didn't even really listen to the song then. In fact, I almost forgot about it. It wasn't until days later, when I was back at home in Berkeley, that I really got a chance to check it out. Of course, I was blown away. It's an incredible piece of music: so warm, deep, and wise."

Dewey wrote the new tune, "GJ," as a gift to Joshua's infant son. It begins and ends with Dewey, alone on alto sax, playing a burst of notes, a downward arpeggio that repeats throughout the song. There is beauty and emotion embedded within the music, expressing the love that Dewey Redman felt for his son and grandson.

"GJ" was Dewey Redman's final recording; he died a few months later. "It was the last time we played together," Joshua said, "and it was the last time I saw him before he passed away. It was an incredible experience to play with him, and neither of us realized the significance at the time; that was May and he passed away in September. For me it was tremendously significant that I had the great Dewey Redman on my album, and that I had a chance to play with my father again."

Chapter 7

CREATIVE PARTNERS

Over the years in public radio, I've become adept at the art of the pitch, and have taken part in fund drives of various tones: some offer earnest, rehearsed, and studied reasoning with listeners, others emphasized pleading. One of the first drives I worked on included threats from Steve Post, the brilliant and wry WNYC morning music host, to keep playing Pachelbel's Canon—over and over and over again—until listeners anted up.

When I was a nighttime music host at WNYC in 1990, the drives were relaxed and playful. They were produced by Mark Maben, who introduced elements like the "Gorilla Challenge," where a taped drum roll would play under the conversation of the hosts. He wore silly hats and made sure that everyone had lots of bagels to keep them going. Fund-raising is a tense time at any station, but his approach infused the chore with a liveliness that energized the staff and actually entertained the listeners. During his tenure, Mark raised more than $1,000,000 for the first time at WNYC during a single fund-raiser.

I met Mark when we pitched for *Prairie Home Companion* together. Between fund-raisers, he was in charge of on-air promotion, and a couple of weeks after our first pitch, we met in Studio Five to talk about promoting my evening show, *One Music.* Mark arrived with an idea about how classics can come in many styles, from symphonies to jazz to folk music from around the world. I loved the direction but wanted to see if we could spin it a little.

"What if," I wondered, "the copy talked about other kinds of classics, like cars, or fashion, or books, while the music behind it showed the range of the show?" Mark was intrigued, and worked on revisions. A few days later, we met again; his new copy was fantastic, fun, and unexpected, juxtaposing little black dresses and peasant blouses, Picasso and Rembrandt. We recorded the tracks, Mark chose the music, and when the promos were finished he played them for me. I still think they're some of the best radio work I've ever done.

They certainly resulted in the best creative partnership I've ever enjoyed. Our work together at the station led to dinner and movie dates, and about two years after we met in the studio, Mark and I were married in a very DIY ceremony in a light-filled photographer's loft just outside the entrance to the Lincoln Tunnel in New York. Our friends brought in a forest of flowering branches from the Flower District and hung them from the ceiling to create a bower; other friends sang madrigals; one played the harp; my childhood neighbors made my veil and the covering for our chuppah; and Mark and I wrote our own vows. It was a beautiful day.

I knew from our first radio collaboration that Mark possessed a creative spark that made work fun, and led to innovative and successful results. And I guessed that this quality would make him an excellent partner in life, too, which has certainly been true. What I couldn't know then was how essential creativity would become during some of our most difficult times.

A few years ago, our younger son became seriously ill. He was hospitalized just after he turned seven years old, and for three weeks Mark and I took turns sleeping in our son's hospital room and caring for his ten-year-old brother at home. Our son just kept getting sicker, and the doctors urged us to consider either serious and life-altering surgery or a relatively new medical treatment with unknown long-term side effects. After hurried, anxious talks in the hospital lounge, Mark and I made the heart-wrenching decision to choose the medical approach, and we rejoiced when it worked. Our son was soon able to come home.

But within a year the medication stopped working and he fell seriously ill once more. Surgery again was advised. Mark and I frantically researched what else might be done, and discovered innovative approaches to the illness that involved diet, acupuncture, and other alternative treatments. The doctor said we had a few months to try these different interventions, but he wasn't very confident that anything other than surgery would work.

It was an awful time. We were exhausted, and there were moments when we were afraid we had made the wrong choices. Yet to our great relief our son has made a slow and steady recovery, without surgery. At his recent checkup, four years since he first fell ill, his doctor was surprised and pleased to note that all the indicators of illness were gone. It's hard to imagine this outcome had I not been able to say to Mark, in the midst of a crisis, "What if we tried this?" knowing that he would throw his entire creative self into figuring out how to make it work.

In *Studio 360*, all sorts of artists have talked about the pleasures and challenges of creative and sometimes personal partnerships. We'll tag along on an artistic "first date" with Robert Plant and Alison Krauss, two very different musicians who are each icons in their respective solo careers. Then a couple of longtime friends, the film director Ang Lee and screenwriter and producer James Schamus, discuss a creative partnership that has led to some of the most popular movies of the late twentieth and early twenty-first centuries. And, finally, the influential architects

Robert Venturi and Denise Scott Brown, who have lived and worked together for more than forty years, talk openly about both the successes and the frustrations of a long creative marriage and working partnership.

In each of these particular collaborations, one partner is American, the other was born somewhere else. Each pair explores iconic American landscapes: bluegrass music; cowboys; suburbia. The combination of an insider's and an outsider's experience adds complexity and richness to the work they create together.

Musician Alison Krauss

"The whole thing felt new to me, and wrapping your head around making a duet record that is not about the two of us being together all the time [but] was really celebrating the difference in our voices."

Musician Robert Plant

"I wanted to work with people who were going to push me, and . . . challenge my whole capacity to be a really proper singer."

Playful Synergy

In 2009, the album *Raising Sand* swept the Grammy Awards, receiving prizes in categories as disparate as Country, Pop, and Contemporary Folk/Americana Album of the Year, as well as grabbing top honors for Album of the Year and Record of the Year. *Raising Sand* was the fruit of a new collaboration between two musicians who had performed together only once before, and who arrived at a Nashville recording studio from very different musical traditions.

Their musical "courtship" began with a phone call, when

Robert Plant, who helped define rock and roll in the 1960s and seventies with Led Zeppelin, dialed up Alison Krauss, best known for her extraordinary bluegrass fiddling and her ethereal soprano. Plant said that he had been searching for a new musical adventure, and a friend suggested he contact Alison because Plant loved the bluegrass recordings she'd made with her band Union Station. "The whole idea of picking up the phone and talking to a complete stranger about a possibility is pretty random, it's pretty nuts, really," Plant said. "Especially as talk is so cheap in our game."

But he made the call, which Krauss said she answered with a quiet, whispered, "Hey, what's going on?" because she was lying next to her infant son, trying to put him to sleep. "It was about seven o'clock, and I had to keep my voice down and talk real monotone so I wouldn't wake the baby up, but of course [Robert] had no idea what was going on," Krauss said. "And he says, 'Well, let me give you my number!' And I said, 'I can't really get up . . .' and he says, 'Well, all right!' And I figured that was the end of my conversations with him for the rest of my existence. But I was very thrilled to get the call; he was complimentary of what our band does, and I really thought it would probably end there."

Plant did call back a couple of years later, and convinced Krauss to join him in a tribute to the great blues musician Leadbelly, who had been a huge influence on both the rock and roll idol and the bluegrass prodigy. Krauss later said, "The minute I met Robert, I thought, 'Oh my goodness, this will be fun.' . . . I walked in the room and saw this big pile of hairdo and walked that way, and he turned around, and he had his glasses on, and said, 'Ahh, there you are.' We talked about [bluegrass legend] Ralph Stanley and traveling through the Appalachian Mountains and how much he loved traditional music." To which Plant, mindful of his reputation as a rock star, slyly added, "I just wanted to touch your leg," and Krauss quickly explained that he was "just kidding about that. . . ."

Robert Plant rocked the sixties as the slender, bare-chested lead singer for Led Zeppelin. His raw and plaintive voice turned songs like "Stairway to Heaven" and "Whole Lotta Love"—and the singer himself—into icons. After drummer John Bonham died, the band broke up in 1980. Over the next twenty-five years Plant wrote and recorded award-winning solo albums, and collaborated with old friends like Jimmy Page as well as with musicians from North Africa.

The year Plant left Led Zeppelin, Alison Krauss was nine years old, growing up in southern Illinois, and already a phenomenal bluegrass fiddler. Krauss was signed by a label at fourteen, and since then she's sold more than eight million records and won twenty-six Grammy awards, more than any other female artist. "My mother always said, 'Don't plan on being an artist and then study to be a secretary as something to fall back on, because you'll fall back on what you study to be.' So she and my dad encouraged my brother and me to follow what we wanted to."

At this point in the *Studio 360* interview, Plant interrupted Krauss to say, "I'm also pleased to say that Ms. Krauss has now stopped flinching when I walk past her." In conversation, Plant and Krauss tease each other like old friends. "I saw a review in an English paper, it was a fantastic review," Plant told Kurt. "But it said, 'Not long ago, people would've been quite concerned about Ms. Krauss's well-being working with . . .'" Krauss quickly interjected that Plant had been a gentleman throughout their collaboration, to which he archly commented, "Oh, honey," and she said, "Oh, c'mon. Oh, *nice*! Now he's raising his eyebrows."

The playfulness in their relationship comes through in their music as well, perhaps nowhere better than on the Everly Brothers's classic kiss-off tune "Gone, Gone, Gone." In the video for the song, Krauss sashays onstage in a flouncy black dress and high-heeled boots, passing through shimmering silver threads onto a set hung with disco balls, her long blond hair flowing. Then Plant saunters on, shirt buttoned up to his goatee, his mane of golden ringlets bouncing to the propulsive beat of drums and guitar. As soon as they catch a glimpse of each other they smile—a genuine "I'm having fun!" smile. Krauss is beautiful, and while at times she tries to look sultry, she can't hold it for long before breaking into a grin. Part of why this song is such a kick is that it's clear that both singers are enjoying themselves. It's no surprise that this was the first hit of the album, winning a Grammy for Pop Collaboration of the Year.

There's a wonderful versatility to their partnership as well. Krauss sings the melody on "Gone, Gone, Gone," with Plant swooping around her on harmony. For other songs, like the 1950s "Rich Woman," Plant carries the tune and Krauss shimmers above and below. Their voices, while very different, blend beautifully.

Another key to the success of their collaboration—and the record—is the brilliant guitar playing of producer T-Bone Burnett. In "Gone, Gone, Gone," Burnett appears onstage midway through the video, standing between the two singers, his long blond bangs flopping as he plays guitar. "In a way, it's almost like a kind of revue where sometimes a guitar part is more important than the vocal," Plant said. "Or the actual rhythm and the way that the percussion is alluring and seductive—it's like a moonlit night in the suburbs of Savannah. . . . It was very amorphous and animalistic. At times, the synergy between people was spectacular."

Krauss had worked with Burnett initially on the music for the film *O Brother, Where Art Thou?* "I'd been around him enough to see that every time that I got in the studio with him, I was very inspired. I loved the tunes he picked out for me, and I loved the way he ran the sessions. . . . I thought he would be a very interesting person to work with. He has such a love for the traditional music."

"Undoubtedly, T-Bone is out there," Plant added. "He's got so many things in his head—it's like a revivalist meeting just hanging out with him. It's spectacular. But more than anything else, he's got such a love of music. And his guitar playing on this record creates a sultry, beautiful, sensual moment on several of the tracks in the simplest fashion across all this color. He comes in and just by touching the root-note strings of the guitar, it shimmers into some other place, which is where I want to be."

For *Raising Sand*, Plant and Krauss covered bluegrass staples like "Your Long Journey," by Doc Watson and his wife, Rosa Lee. The mood is somber in this song about death, and that feeling infuses much of the album. As they chose the songs to perform, Krauss insisted that they explore dark material, because she felt that the convergence of their voices took them into new territory, a place tinged with sorrow and loss. "There's so much life and experience that his voice brings out, there's a lot of mystery to it, and with mine, together, that creates some kind of story, and I don't think it creates an 'up' story. I think it creates a lot of wonder, and it creates sadness. That's the emotion I feel when I hear us singing together. It's something that has a past."

The record includes moody songs from the fifties and sixties, and country tunes from the seventies. "Boy, it felt all new to me," Krauss said about the process of recording with Plant. "It was very different, very free, it wasn't as meticulous as we are when we go in the studio [with Union Station]." The songs have haunting, fluid rhythms, with Krauss and Plant often hovering around the beat, one coming in just ahead, the other pulled along just behind. T-Bone Burnett described the rhythms they played with less as a beat, and more as a wave. "T-Bone was very much into what happens at that moment when he produces," Krauss said. "I loved the process, which very much involved everybody. . . . The whole thing felt new to me, and wrapping your head around making a duet record that is not about the two of us being together all the time [but] was really celebrating the difference in our voices. It's very different in bluegrass, which is trying to sound as one all the time. It really shows the differences in our voices, which made it romantic."

Robert Plant is delighted that he called Alison Krauss all those years ago; in 2008 he chose performing with Krauss on the road over reuniting with his remaining Led Zeppelin bandmates for what would have been a lucrative tour. "Something had to happen for me where I was going to learn something," Plant said. "I wanted to work with people who were going to push me, and not threaten me but challenge my whole capacity to be a really proper singer. Not just a one-trick pony, but somebody who could actually modify and adapt and get into some kind of dreamscape. As it happens, this is the combination that transpired, and it couldn't be better, really."

Director Ang Lee

"I think that's how those films work;
they're very specific, but there are universal things
that everybody in the world can relate to."

Screenwriter and producer James Schamus

"It's astonishing to me . . . having helped on screenplays
for Ang's pictures . . . that weird divide that happens
when you feel such ownership to what you've
written on the page and then you hand it over."

Crossing Cultures

In 1991, an aspiring filmmaker scheduled a meeting with producer James Schamus at his New York production company, Good Machine. "He came into the office," Schamus recalled ten years later, "and he said, 'I'm Ang Lee. You probably don't know me, but if I don't make a movie I'm going to die.'" Lee described Schamus at the time as a "mixture of a professor and a used car salesman" sitting at one of the two tables in Good Machine's bare-bones office.

From that initial encounter, a long-term creative collaboration was born. James Schamus has written or produced every one of the films that Ang Lee has directed, award-winning movies that have a tremendous range of time and place, including *Brokeback Mountain*; *The Ice Storm*; *The Hulk*; *Crouching Tiger, Hidden Dragon*; and *Sense and Sensibility*.

Before meeting Schamus, Lee had made well-regarded films as a student at NYU, where he received his master's degree. But after graduation, he spent six years in what he calls "preproduction hell," writing and rewriting screenplays and taking care of his two young sons while his wife, a microbiologist, worked to support their family. Finally, he won a Taiwanese screenwriting contest, which allowed him to shoot his first movie. That's when he walked into Schamus's office, and together they made *Pushing Hands*, a film in Chinese about an older man from Taiwan who comes to the United States to live with his son, daughter-in-law, and grandson. Audiences in Taiwan loved it.

In his next film, Lee continued his exploration of family and culture, writing and directing *The Wedding Banquet* about a gay Chinese man named Wei Tung who lives in New York with his American lover. To keep his parents, who are anxious for a wedding and grandchildren, off his back, he engages in what he thinks will be a sham marriage with a young Chinese immigrant so that she can get a green card. When Wei Tung's parents arrive in New York, expecting a big celebration, the strain of keeping up an elaborate romantic charade leads to amusing and poignant complications.

The Wedding Banquet was an unexpected hit. "It's also the first movie James significantly touched up for me," Lee said. Schamus added that working on the screenplay was a lot of fun, "because the original draft, believe it or not, was not funny. It was a tragedy. When the parents discovered that their gay son has married this young Chinese woman to hide his identity from them, they have a gigantic fight, everyone screams at each other, and they go home in tears. That was the [original] end of the movie."

In his rewrite, Schamus mined the comedic aspects of cross-cultural miscommunication and intentional misdirection, as well as the deeper implications of trying to please one's parents, even when it requires pretending to be someone else. Midway through the movie, Wei Tung and his bride get married in a dingy city office. The scene has roots in real life—Schamus and his wife were married at the Municipal Building in New York; Ang Lee married his wife there, as well. Schamus says he wrote an initial draft of the scene that Lee found charming, but Lee's emphatic note was that it wasn't funny enough. So Schamus pushed the humor, and in the final cut, the bride mangles every single line that the justice of the peace feeds to her—ending with a forceful "in sickness and death," while omitting the health and happiness part.

After the wedding vows, Wei Tung's lover, Simon, takes a photograph of the couple with Wei Tung's parents, and his mother bursts into tears. Lee said that this scene in *Wedding Banquet* was like a documentary of his own wedding, when his mother just kept crying and saying, "We're useless! This is such a shoddy disgrace!"

"*Wedding Banquet* presented to all of us this opportunity to take a topic that was really serious and really important," Schamus said, "and use an American genre, in this case, the screwball comedy of marriage. Just making it gay and Chinese kind of revived it, and we were able to use the exact formula that Howard Hawks, and George Cukor, all those guys, used, but going straight to how gay it was even back then. They were just pretending it was straight. And suddenly it was kind of magic." Lee and Schamus made the movie on a tiny budget, and it became a huge international success.

This longstanding and productive partnership is a study in contrasts. Both men grew up in the 1960s, half a world apart. Ang Lee spent his childhood in Taiwan; his father had fled there from mainland China, where Lee's paternal grandparents were executed during the civil war. Lee said he was born to be an entertainer—and that this was very difficult for his father, a high school principal, to accept. Because Lee was a first son, his father wanted him to study and go to college, to carry on the Lee name, while all Ang wanted to do was go to the movies. "I always loved watching movies. And I'm . . . very easy to cry. Sometimes I cried so hard the whole row of audience laughed, 'Look at that kid!' I grew up not being encouraged to distract from academic study, do the right thing. So I daydream and I never think I'll be a filmmaker."

In high school, Lee appeared in amateur theater productions that extolled the virtues of the anti-Communist regime, often playing the leading man. "Kind of pale, good-looking, I believe. Usually you have to be good-looking to be the hero. And patriotic, and have the right thinking. Never touch the girl. They just fall in love with you but you don't touch them." He went on to study acting in college in Taiwan, then came to the States to study at the University of Illinois. His poor English meant that acting was out of the question, he said, though he did have a short stint as a mime. Theater directing didn't appeal to him; he wanted to be a film director.

While Lee was watching movies in Taipei, James Schamus was sitting in dark theaters in Southern California. "Movies were always a passion, and I think, the same as Ang, I grew up watching a lot of crappy movies," Schamus said. "Anything that was out there. And the joy of those moments that you connect to the cinema means you carry that with you your whole life. So you can't be a snob. There are great moments when mass culture really works. But I also had moments, growing up during that time in L.A., of public television and the local critic showing films on Friday night like *Birth of a Nation*, and *Intolerance*, and *Sunrise*, and great silent black-and-white films. By the time I was nine or ten I got hooked on that stuff, too. So it was a strange concatenation."

In college, and later in graduate school at UC Berkeley, Schamus focused on literature; movies took a backseat. He started off "as a Miltonist, studying the English Renaissance, and slowly shifted into American [literature], and then photography. And then suddenly this love of movies became more present. It was a great time to germinate and let that stuff fizzle and to take my time. By the time I landed in New York and started working as a production assistant, I was the oldest living PA in the history of the film business of New York."

Both men found tremendous inspiration in the films of a certain Swedish director. "[Ingmar] Bergman is a bonding thing for us," Schamus said. "As a graduate student, I taught Bergman. My [academic] specialization is, in fact, Scandinavian film." Ang Lee said that it was a Bergman film he saw when he was eighteen that inspired him to dream about becoming a director. "I was actually in Academy of Art and I watched Bergman's *Virgin Spring*. That was an epiphany for me. The director is a superstar. How wonderful, if you can ask where God is with such beauty. I remember that. That sort of turned my life around."

Just before Bergman's death, Lee and Schamus were invited to show one of their movies on Bergman's island off the coast of Sweden. "They had organized some screenings and he had selected *The Ice Storm* as the one non-Bergman film they screened that summer," Schamus said. "And the woman running the screening said, 'Please come out and visit, but know that Bergman is very aged'—he was eighty-eight, eighty-nine then—'and he probably won't come out of his compound. Don't expect to see the guy.' In fact, they organized the dinner and he didn't show up. And then we got the phone call the next afternoon saying, 'If you can be ready in ten minutes, the car will pick you up. Come on out and meet the man; he's actually invited you to his house.' The scene was just extraordinary when Ang got out of the car and saw him."

"A hug," Lee said. "It was like this man took away my innocence, so to speak, when I was eighteen. And now he gives me a hug, like a very motherly hug. It was something. Life is all right, it's a pilgrimage experience for me."

Going to the movies and watching TV are recurring themes throughout Lee and Schamus's films. In *Brokeback Mountain*, Jake Gyllenhaal's character, Jack Twist, finally stands up to his father-in-law over whether or not to watch TV during Thanksgiving dinner. Christina Ricci and Elijah Wood spend hours in a typical suburban den in *The Ice Storm*, watching TV and awkwardly making out. And movies play a central role in Lee's 2007 thriller *Lust/Caution*, set in Shanghai in the 1940s, when the Japanese occupied the city. The main character, a rather naïve young woman from the provinces named Wong Chia Chi, tries to escape the long food lines and dying bodies that line the streets by ducking into a glamorous movie theater, which is showing the 1941 Cary Grant–Irene Dunne film *Penny Serenade*. Wong Chia Chi loses herself in the melodramatic fantasy, only to have the picture interrupted by Japanese propaganda. As the young woman leaves the theater, an encounter with an old school friend thrusts her into a world of deception, sex, and murder worthy of any classic thriller.

"The story on which we based the movie, written by one of the great writers of the twentieth century, Eileen Chang, is a story about women in particular who go to Hitchcock movies and end up finding themselves in them," Schamus said. "That's really the biggest twist of all. When you start to take your reality and bridge it and mold it and craft your own identity in response to what you've seen on the screen. And so, in finding yourself in that film, the film certainly has a lot of those moments and twists, but more importantly the biggest one of all is—[once] you find yourself inside that screen, what do you do?"

"I found that the movie—Hitchcock is the last one to do this, the old-fashioned film noir—is romantic," Lee said. "It's really a following to the film noirs in the forties, and combined with melodrama. So I think when we talk about Hitchcock, we're not obeying the genre but rather mixing it in."

Listening to Schamus and Lee talk about creating this movie offers a bit of insight into how these longtime creative partners collaborate. They spent a couple of years working on the screenplay with a cowriter, Wei Ling, who is an expert on Eileen Chang. "My purpose and role in this was to shake it loose from the story, the underlying story, while maintaining the spirit and integrity of the vision," Schamus said. "So for me it was fascinating. But more important, it was trying to think cinematically. Break out of both the history and underlying material—just a bit—to give it room for Ang's maneuvers on-screen. To make things happen. For example, the picture really bends itself in half. There's a moment halfway in the picture, when all these silly young students who are trying to become resistance heroes are faced with having to perform a real murder. In the story, the end of that first half, which takes place in Hong Kong before they scattered, dispersed, and refound each other in Shanghai a few years later to continue their quest to kill this traitor, it's more a natural scattering. They fail, and then there's the Japanese attack on Hong Kong, everybody disperses. What we tried to do was come up with something that was much more cinematic, and in the mold of the movies we were paying homage to, and hence this very extraordinary murder scene. One of the most memorable I've ever seen on-screen . . . What we try to do is multiply the points of view and make the murderers themselves feel the same sense of threat coming their way."

The scene is visceral, bloody, and horrifying. It takes place just after the heroine loses her virginity, in preparation for becoming the traitor's mistress. "And therefore," Lee said, "the boys should lose theirs. I call the scene 'Bar Mitzvah.' I think James wants to, in the middle part of the movie, wake up the audience. But when I see it and I read the first draft of that scene, that's the word that hit me—'Bar Mitzvah'—for the boys. That's a great device to separate the movie into two halves. A perfect device in that it's not just the excitement, [but also] the disillusion."

Schamus and his colleagues at Focus Films decided not to pursue a rating for *Lust/Caution*, which is unusual. Ang Lee said he knew from the beginning that the sexual relationship between the illicit lovers was central to their characters' discovering who they really are. "The title has 'Lust'—to me, that's the anchor, and that's the ultimate performance, too. I shot those three major sex scenes relatively early to the five-month shooting schedule. Only when I know how it's ended, I will learn how to craft the second part of the movie. So, yes, they were very important to me. I wasn't thinking about rating, I just do the best I can, and bring the two brave actors along. No one knows how it was going to go ahead. Including myself."

"The fact is, the film is very much about truth: emotional truth, narrative truth," Schamus said. "Who's telling the truth? What's real? And here are two characters—one is a spy, the other is spying on that spy, seducing him, with the purpose to assassinate him. . . . Sex is one of those spaces where everything is absolutely true or everything could be completely fake. So performance is required to get to a certain kind of truth. They had to be fake in order to find something else that was quite true."

Hiding one's true identity, telling lies about yourself in order to make your way in the world, are recurring themes in Lee and Schamus's work, whether it is a gay son who gets married to please his traditional parents in the *Wedding Banquet*, or a proper young nineteenth-century Englishwoman in *Sense and Sensibility* pretending not to care that her beloved is marrying someone else. When Kurt asked Lee why sexually charged identity crises were a favorite subject, he answered, "I wish I knew. I guess maybe, first of all, I'm sexually repressed. Secondly . . ."

Kurt interrupted Lee here with a skeptical, drawn out "Okay . . ." And James Schamus responded, "It's a good line to use!"

It's almost exactly the line that Lee himself delivers in *The Wedding Banquet*; he makes a cameo appearance as a guest who watches the over-the-top shenanigans of the bridal party during the elaborate banquet. A Western guest sitting nearby remarks, "I thought the Chinese were meek and quiet math whizzes," to which Lee's character replies, "You're witnessing the results of five thousand years of sexual repression."

"I do have identity problems," Lee said. "What's the deepest cultural root for me is the classic Chinese culture, which is whimsical for me. It's like a dream. It was taught, passed down by my parents. When they left China, they romanticized it and taught it to me. That's my cultural root, which is diminishing and changed in all parts of China.

"All my life I've been a foreigner. I've been an outsider in Taiwan, and then I come here to the States as a foreigner. When I go back to China, I'm Taiwanese, or American, whatever. So in some ways I'm orthodox Chinese, but in other ways I'm like an outsider. Because I'm not really rooted in that culture, my resources come from reality, not theatrical conventions."

This perpetual outsider is able to create films that exquisitely penetrate the ambience and mores of so many different times and places, from the deep divisions within a Taiwanese family in *Eat Drink Man Woman*, to the two cowboys' experience of riding horses, tending sheep, and falling in love in *Brokeback Mountain*.

As a screenwriter, James Schamus has had to immerse himself in Chinese culture in order to create compelling stories for many of Lee's films, including *Lust/Caution* and *Crouching Tiger, Hidden Dragon*. Schamus and Lee have talked about how much of a challenge it was for Schamus to get the right tone for the interactions of a Taiwanese father and his three daughters as they wrestle with the changes that are taking place in modern Taipei for *Eat Drink Man Woman*. "Everyone always tells young writers 'Write about what you know,'" Schamus said, "so to find yourself as an American writing about a Taiwanese family when you've only spent a few hours in Taipei pretty much breaks that advice down to the farthest reaches of uselessness. So I did a lot of research, and the more research I did . . . the worse the script got."

"He tried very hard to be a Chinese," Lee responded, "but it just doesn't read like Chinese. So I kept complaining until he was so frustrated." Schamus's solution? "I made a global change in all the [Chinese] names in the computer, as I was writing the script, to Jewish names, and just let it rip. Because it's about food, it's about family—that's kind of Jewish." Lee was thrilled with the results: "I said, 'Oh, it looks very Chinese! Good improvement!'"

"It's astonishing to me over the years, having helped on screenplays for Ang's pictures, both in English and in Chinese, that weird divide that happens when you feel such ownership to what you've written on the page and then you hand it over," Schamus said. "And then you see Ang and other other writers struggle with the translation and modification to make it culturally appropriate, to make it work verbally. And understand at the end of the day that I really only understand half of what I've written. Because as it comes back and is transformed in Ang's hands, so much else happens."

Whether set in bamboo forests in China, suburban Connecticut living rooms, or pastures in the Rockies, Ang Lee's movies connect with audiences all over the world. Lee mused about why these stories cross cultural boundaries so easily. "It must be something universal among all of us. I think that's how those films work; they're very specific, but there are universal things that everybody in the world can relate to. The beauty to me of working with James in those projects is bringing those textures."

What also unites the movies he's chosen to direct, Lee told us, is that he has no interest in straightforward stories. "They have to be somehow twisted. Dramatic. Have enough elements in conflict with each other. That seems to make sense to me, even to a degree of frightening [me]. That would interest me—to go a year or two of my life into it. And probably at the end of the process I still couldn't quite figure out how to pass on the enigma to the audience and see how they respond. That's the way I learn about the world and myself."

Architect and urban planner Denise Scott Brown

"Being a married couple and partners in work produces maximum possibility for abrasion."

Architect Robert Venturi

"I think it's one of the luckiest, best things of my life—working together and coming together."

Intertwined Creativity

If someone made a movie about the celebrated and sometimes controversial architects Robert Venturi and Denise Scott Brown, the first scene might be set in a contentious faculty meeting at the University of Pennsylvania in 1960, held to discuss what to do with the 1891 library designed by the legendary Philadelphia architect Frank Furness. The library featured a complicated brick and terra-cotta façade, intricate leaded windows, and a light-filled, vaulted reading room. When it was built, it had been widely acclaimed as a masterpiece. But by the time of the faculty meeting, sleek modernism was in fashion and the decorative Furness library was considered an eyesore.

Denise Scott Brown had recently finished her graduate work in urban planning at the university and was a new instructor; this was her first faculty meeting. Robert Venturi, who had worked as an assistant to the great modernist Louis Kahn, was teaching architecture classes, and it was one of his first meetings at Penn, too. "The main question for that faculty meeting," Robert Venturi said, was "should the school of design take a stand concerning the demolition of this great Victorian building? How could there be a question about that? Of course the school should take a stand and say no. But at the time, Victorian architecture was not accepted by architects. Denise made a beautiful plea. She said, 'Yes, we must take a stand, this building should not come down!' Afterwards I went up to her, I had seen her around a lot, of course, but we had not met. I said, 'I want to introduce myself, my name is Robert Venturi, and I want to tell you how much I agree with what you just said.' And her first words to me were, 'Why didn't you say something?' And it's kind of been that way ever since."

That encounter led to a creative partnership that has spanned fifty years, during which Robert Venturi and Denise Scott Brown have worked together as teachers, writers, and designers, as well as marrying in 1967 and raising a son together. Denise Scott Brown agrees that their interactions have not changed very much since that first meeting. "Being a married couple and partners in work produces maximum possibility for abrasion," she said. "Yet when we argue, it's usually more to do with our lives as people married to each other than with our creative collaboration in architecture. We fundamentally trust each other about architecture. But, sadly, our joint creativity in design is ignored or denied in the profession because most architects seem to feel that great architecture must emerge from one brain only."

Together, Robert Venturi and Denise Scott Brown have designed and planned hundreds of buildings and college campuses around the world, both delighting and dismaying the architectural establishment with their projects, including an addition to London's National Gallery, a gas station for Disneyland, and a master plan for the University of Michigan. They've written some of the most influential architecture books of the last fifty years, including *Complexity and Contradiction in Architecture* and *Learning from Las Vegas*. From the beginning, Venturi and Scott Brown have been completely comfortable questioning the modernist architectural orthodoxy of their time, and their innovative ideas and passion for opening up the vocabulary of contemporary architecture to everything from classical columns to highway signs have positioned them at the center of important and interesting controversy.

You may never have seen one of their buildings, but you've seen their influence. Kurt describes their role in disrupting the geometric, some would say severe, forms of modernist architecture this way: "Modernism really limited the palette of what was acceptable, what was legitimate, in architecture. It was as if you'd said to painters, you can use only black and white and a variety of grays. Of course, at the same time, people were building houses and strip malls that didn't conform to the Modern Orthodoxy at all, but for serious architects in the Modern tradition the palette was really limited. Then Venturi and Scott Brown said, 'This is ridiculous, we have hundreds and thousands of ways of doing architecture, why limit ourselves? Let's use color, we can stick a column in the "wrong place," or design a pitched roof.' We could do things that Modernists had taken off the table in their zealousness to declare new architecture for a new world."

Venturi and Scott Brown arrived at the University of Pennsylvania from intellectually similar, yet geographically different, paths. Robert Venturi grew up in Philadelphia, where his grandfather and father were produce merchants and his mother was an ardent socialist who sent her son to a Quaker school. Neither of his parents was able to finish high school because they needed to go to work. Both loved architecture, and on drives through the city they often pointed out interesting buildings to their son. In his first book, *Complexity and Contradiction in Architecture*, Venturi remembered how much he hated the fussiness of Frank Furness's Victorian buildings when he glimpsed them during those trips. It took many years before he began to admire Furness's exuberant ornamentation and use of industrial materials like brick and steel.

Venturi traveled across the Delaware to Princeton University for college, where he earned a B.A. in architecture and a master's in fine arts. During his graduate studies, he spent a summer in Europe and, later, two years in Italy as a fellow at the American Academy in Rome. He became captivated by baroque architecture in England, explored modernist masterpieces by Le Corbusier in France, and fell in love with the classical architecture of Rome.

Venturi often includes classical references, mixed with a modern sensibility and an unexpected approach, in his designs and his writing. An excellent example is the house he completed in 1964 for his mother, set back from the street in the leafy Philadelphia suburb of Chestnut Hill. Venturi put all of the major living spaces on one floor, with space for a maid or a nurse above. There's an iconic photo of the house with Venturi's elderly mom sitting in the doorway of what one critic described as an oversized child's drawing of what a home might look like. To picture it, think of a single-story ranch house stretched out over its lot, with a giant pitched roof that covers the entire long rectangle. Now imagine that the center of the huge triangle is split, so that you can see through the pediment of the roof to a large chimney (which also houses a stairway) in the back, and that the gap is connected by a symbolic arch, also split down the middle, over the doorway. "The house is big as well as little," Venturi wrote in *Complexity and*

Contradiction, "by which I mean that it is a little house with big scale. Inside the elements are big: the fireplace is 'too big' and the mantel "too high" for the size of the room; doors are wide, the chair rail high."

Venturi's delight in playing with historical elements in a modern building was unusual at this moment in architectural history, and his mother's house received both passionate critical enthusiasm and vitriolic outrage. "I think it's important that when we realized historical architecture was valid to study, we said use it as a reference. We didn't say re-create it," Venturi pointed out. "Reference to the historical past can enrich architecture; copying stultifies it."

In addition to his passion for history and modernism, Venturi's later work also reflects a deep fascination with popular culture, something he hadn't really focused on until he met his wife. "Denise Scott Brown got me into the pop culture. I think that, in the early days, she, as a foreigner, could see the American culture more vividly and more objectively than I, who had grown up within it."

Scott Brown was born Denise Lakofski in Africa, where her grandparents had immigrated from Latvia and Lithuania. Her father was a businessman in Johannesburg, her mother had studied architecture at the University of Witwatersrand, and together they built a home in the International Style, where their daughter played on the flat roof.

When she was five years old, Denise decided to become an architect, and later followed her mother's path to the University of Witwatersrand's program in architecture. In 1952, she transferred to the Architectural Association School in London and became deeply immersed in the social aspects of architecture. "In England, I found similar disparities [to South Africa] but they were based on class rather than colonialism. They existed between, for example, working-class street life in London's East End and middle-class suburban life in the new towns—where rebuilding plans decreed bombed Londoners should be moved—to the 'country,' where it was healthy."

With her first husband, Robert Scott Brown, she applied to the University of Pennsylvania to study city planning. "My African and English experiences made me agree with American sociologists when, in the 1960s, they complained that urban planners' class values were dictating how all groups should live in the renewed city." Not long before she met Venturi at the faculty meeting, Scott Brown's husband was killed in an automobile accident. She returned to Penn to complete masters degrees in architecture and planning, and stayed on to teach.

Venturi and Scott Brown taught classes together, and were close professional colleagues. Then Scott Brown left to teach at Berkeley and UCLA, and decided to stop at a few cities along the way on her trip west. "Visiting Las Vegas in 1965, I thought, 'Venturi needs to see this.' So when I moved to Los Angeles I invited several colleagues to lecture to my students at UCLA. One was my old friend, Bob. After his lecture, I gave the students a sketch design [assignment], and Bob and I spent four days in Las Vegas."

"In some ways," Venturi said, "Las Vegas brought us together." Scott Brown added, "We worked and taught happily together at Penn for several years but perhaps I had to go away, perhaps he had to go away, before we could come together. Who knows? Perhaps Bob had to publish his book to be ready for a different stage in life; perhaps I needed time to grieve the death of my first husband. For whatever reason, we knew each other for seven years before we married."

As they were falling in love in Las Vegas, the city was becoming a catalyst for their creative lives. Venturi says that their initial response to the architecture there was a volatile mix of emotions. "We studied Las Vegas more formally later on, but from the first, we were both horrified and fascinated by it. We used the term 'love/hate' to describe these feelings. We saw, and still see, 1960s Las Vegas as one of the most significant cities in America. Los Angeles is *the* city of the automobile. The Las Vegas of that time was an extreme example. We had reacted against Modern architecture's urban ideals—against Frank Lloyd Wright's belief that Americans should live in middle-class suburban utopias, like his Broad Acre City; and against Le Corbusier's Ville Radieuse, a park with high-rise buildings, to be built by tearing down most of Paris. We were pragmatists, and we felt that sometimes evolution could be more appropriate than revolution."

Inspired by that visit, Scott Brown had an idea for a studio that the couple created at Yale's School of Art and Architecture in 1968, where both were teaching. With a group of mostly architecture students, they examined Las Vegas's landscape, architecture, and culture. Later, they wrote the groundbreaking book *Learning from Las Vegas* with Steven Izenour, their assistant for the studio and later a member of their firm. It's a brilliant study of urban sprawl in the middle of the desert that was both serious and playful, analytic and artistic. The big book is filled with images: photographs of the gaudy hotels and other buildings lining the strip, drawings and doodles of classical buildings in the margins. A pair of images and the ideas they elicited from Venturi and Scott Brown became famous: a nondescript rectangular building with a giant sign on a pole in front, and a store in the shape of a huge duck. For the first image, what the team called a "decorated shed," they wrote that "the sign is more important than the architecture." For the duck, "the building *is* the sign."

"It's important to recognize the context," Venturi said. "I'm simplifying, but for forty or fifty years Modern architecture shunned symbolism and iconography. Its forms were based on abstraction. Yet ironically, the modern architectural vocabulary was also derived from the forms of industry, which it used to symbolize a new world. So when we said, 'It's time now to learn from the American commercial, rather than the American industrial, vernacular,' we meant that our society had gone beyond the industrial age and that its symbolism was outdated. Combining industrial forms with abstraction was boring and no longer made sense. From Las Vegas we learned to look at architecture not only for its formal and spatial qualities, but for its symbolic content. *Learning from Las Vegas* suggests that architects, in the second half of the twentieth century, acknowledge symbolism. In an era of multiculturalism, we called for study of cultures beyond highfalutin high culture, and for including pop culture, which is valid and can be inspiring. Again, there was good historical precedent for doing so."

In his introduction to the book, Venturi went to great lengths to acknowledge that this book was a collaboration, that Denise Scott Brown "is so intertwined in our joint development that it's impossible to define where her thought leaves off and mine begins." Their work together, he told us, was a natural evolution. "It wasn't a dramatic decision. It just happened. Circumstance brought us together and then we worked beautifully together."

In the early years after they married, Scott Brown said, "Bob was running a small practice that lurched from emergency to emergency, and I found myself working in the office because they needed me—I knew about urban planning, for one thing, and could help the office earn more money, but mainly we had a good architectural collaboration. After spending a year helping out at both Yale and the office, I said, 'I'm working here full-time, let's make it formal.' So although I joined the office in 1967, when we married, I became a partner in 1969."

Today, at Venturi, Scott Brown and Associates, they share the workload with three additional principal architects, and the firm employs more than twenty people. "It's not that one is a designer and the other runs the firm, nor are we pigeonholed, one in planning, the other in architecture," Scott Brown said. "We share roles. In formal terms, Bob is principal in charge of architecture and I of urban and campus planning and urban design. Although I spend more time on architecture than he does on planning, he has a sense of urbanism that's rare among architects and he makes creative contributions to my projects. I'm an architect and a planner, I love the range. In several recent projects, we started as campus planners, then carried through to a major architectural complex. Conceptual plans for the architecture were evolved by me during the campus planning phases, and, on such projects, I continue to be involved through design development and documentation—even down to the door hinges, because if I don't, spaces that the education plan requires to be public might not be, because their doors can't be left open."

"How we work can best be described by something a college friend, an English professor, told me," Venturi said. "Quoting T. S. Eliot, he pointed out that the creative process consists enormously of criticism. You don't invent all the time. When you get an idea, you try it out, then you critique it. You work much of the time as a critic of your own ideas. I think Denise and I are very much critics of our own and each other's ideas."

"We jump-start the design process," Scott Brown responded, "by batting ideas around between us. Our ideas bounce back and forth. We cap each other's ideas. Sometimes we argue. I say, 'I don't think it's working this way.' He replies, 'Of course it's working.' But I reply, 'Couldn't you consider such and such, do it this way?' And he, 'No, no, you can't do that.' Then, a day or so later he'll say, 'Remember what you said? Well, look what I did.' This questioning intensifies the process and speeds it up. Self-criticism might eventually have led him to the same position, but my questions and suggestions push him and he gets there faster. Of course, he does the same for me."

In conversation and in their writing, the couple emphasize the collaborative nature of their creativity. But this idea often meets resistance from the rest of the world. "It's difficult," Scott Brown said, "to get architects to believe that an idea can originate and grow through the contribution of two or more minds—to conceive of the notion of joint creativity. Yet with us, each idea has its antecedents in our shared past. It's hard to say that it originates in one, not the other, because we've been through so much together. We may both think of it at the same time, or one may suggest it and the other see its value and add what makes it important in the design. The original thought may be inappropriate but when capped by a second thought it may become central. Both contributions, plus a ping-pong back and forth, may be needed to make it great. Yet that view of creativity is discounted by the American Institute of Architects in refusing to award their Gold Medal to two people jointly. They have said that an idea can only originate in one mind, and seem to feel that their medal—and respect for design in architecture—will be debased if the recipients are not individuals."

"Absolutely, that's been a problem," Venturi added. "Perhaps two problems: one, 'A female designer isn't as creative as a male,' and, two, 'To be really creative you have to be a lone genius working by inspiration in a romantic setting.' Both views are stereotypes, and neither acknowledges the complexity in our tasks that pushes us to work creatively together. It's relevant that we're not performing artists. If we were, the nature of the collaboration—who is doing what work, how the creative work is shared—might be more obvious."

"In my role as an architect," Scott Brown said, "I am often stereotyped as a helpmate. People say, 'So you're an architect, too?' My response is to point to Bob and say, 'No, he's an architect, too; I'm an architect.' For some reason, when we contest the stereotypes, people get angry. Men—not all, but some—see such remarks as assailing them. And I've heard one woman agree with the AIA about the origin of ideas and another say she would be happy if hers were attributed to her husband."

Robert Venturi and Denise Scott Brown have been nominated several times for the AIA Gold Medal, always together, and the nomination has always been returned unopened. But each has refused to be nominated separately. In 1991, Venturi did accept the Pritzker Prize, perhaps architecture's most prestigious honor, which includes a bronze medallion and a $100,000 award. "It was difficult for both of us," Scott Brown said. "Bob told them you should give it to the two of us and they said no. Jim, our son, attended the prize-giving to support Bob, and I was really happy he did. But I wouldn't go, and I think everyone knew why."

Their son, James Venturi, a technology entrepreneur and filmmaker, is making a documentary film titled *Learning from Bob and Denise.* Scott Brown acknowledges that when James was a child, she and Venturi struggled with the challenges working parents face. "We felt bad when we gave time to the job and bad when we gave time to our child. We walked a narrow line between guilt on either side. Now I've concluded that my career has been an important element in our son's life. It's given him a second quarry of ideas and thoughts and more options for his own career. I feel that mothers who work can add richness to their children's lives. When a mother has a job, it's as if the whole family has a job, and the kids see working as part of the life of a normal human being—male or female."

"'Are we spending enough time with our child?' was a constant question," Venturi added. "It was a difficult balancing act. But I agree with Denise that our son benefited from seeing the people around him dedicated to work and working very hard. I remember at a reception several years ago, I was talking with a young woman who said, 'My husband and I both work. We feel very guilty that we're not with our children enough.' I replied, 'Denise and I had that problem—it was a strain for us as he was growing up,' but I didn't notice Jim standing behind me till he said, 'Dad, you got it all wrong. I saw much too much of you and

Mom when I was growing up.' That was the result of his graciousness and humor, but I think it turned out that the conflict between family and work was not negative."

As a married couple managing a large firm, they are aware of the dangers of slipping into a parental role with their professional team. "They'll kill you if they think you're being parental," Scott Brown said. "They say, 'This isn't a family,' but sometimes you feel it is. Young architects who see Bob and me disagreeing may say, 'If I did such and such I could satisfy both of you, couldn't I?' So in a way we play the parental role. When I see someone working hard and long, I feel tempted to say 'Get a life' or 'Are you sure you're all right?' And every now and then I slip into a mother role, but I feel my way carefully because it may be unwelcome."

They do have a few principles for living and working together. "I've learned we shouldn't talk when we're hungry," Scott Brown said. "I have a tendency to verbalize a lot, to explain things. And sometimes the explanations don't help; it's better to leave things the way they are. Expounding on why so and so did such and such to so and so may be useful to me. That's the way I think. But they don't help Bob much."

"We don't talk business at the end of the day, driving home from the office, or at home, or when we're on vacation," Venturi added. "That's because we're physically and emotionally exhausted. It's better to do it in the office. It's better to relax together and forget, to be a married couple rather than a professional partnership—so part of our life can be less tense; so we can watch TV, do the crossword puzzle."

What might have happened to their careers had they not met at that contentious faculty gathering so long ago? Venturi said he does not think his work would have been as rich without the influence of Scott Brown. "It would have a narrower range. I think it's one of the luckiest, best things of my life—working together and coming together. It's had an enriching effect on the art. Without her it would have been different, not at all as good. I think the quality would have been high but the range would be lacking. It would not include the social aspects of our work, the multiculturalism, which, in a sense, Denise introduced me to. And it would have missed the breadth of approach that comes from taking planning as well as architectural perspectives."

"I think I have had an effect on the designs of many of our buildings, and that without me some things don't get done the way I do them," Scott Brown said. "When this happens, I may fight and say, 'I think it should be this way.' And sometimes I succeed, sometimes I don't. Projects where I've had considerable say are, I think, different from those where I've had less.

"In America, there was a time when there were few senior women architects. And among those, even fewer were American-born. They came from many countries. To build a practice and make a reputation as a woman architect in America was close to impossible. Now there are many American women in positions of seniority and ownership."

"There is an irony, Denise," Venturi interjected, "that in Europe the chances were better in your generation for women to have identity and responsibility as architects than in America." To which Scott Brown replied, "I probably would've had much less chance if I hadn't been part of this firm, and I probably would've continued as an academic."

In the almost fifty years since their initial meeting, Venturi and Scott Brown have changed the world architectural landscape. Their ideas and designs opened up new vistas and vocabularies for buildings and, for good or ill, led to what is now called Postmodernism. There was a time when skyscrapers in New York sported elaborate tops, like the AT&T (now the Sony) Building in New York, with a roofline that looks like a Chippendale highboy. Venturi and Scott Brown's ducks and decorated sheds were everywhere but mostly designed by others. "In the 1980s,"

Scott Brown says, "Bob's mother's house was on the roof of buildings worldwide. Because I was then a planner for downtown Memphis, some architects came to our office to get my take on a project they were proposing for a riverfront site. It was a huge complex, and on top of every building sat Bob's mother's house. In our meeting room was a photograph of the real house, with Bob's mother sitting on a chair in front of it showing how really small it was. I put it alongside their enormous buildings. It was a bittersweet moment—how we would have loved the opportunity they had! But we certainly weren't Postmodernists in that vein, and Bob and I are happy that a younger generation of architects seems to get what we really stand for."

As I write this, Denise Scott Brown and Robert Venturi are seventy-eight and eighty-four, still working on new projects but thinking a lot about how the firm will continue when they are no longer here. Oh, and that Furness building? Venturi and Scott Brown oversaw the renovation in 1991—and the library is now on many lists of the most important buildings in America.

Chapter 8

REWEAVING A SHATTERED WORLD

For the first eight years of the program, *Studio 360* occupied the thirtieth floor in the circular tower on top of the Municipal Building in lower Manhattan. Visitors often gasped at the panoramic city view through our windows; we could see north to the Empire State and Chrysler buildings; west to the Hudson River; and from my office, looking east, I could glimpse the Brooklyn Bridge. When working late at night, I watched planes coming in for landing at LaGuardia Airport in Queens.

Almost always, in the early years, I was the first to arrive at the office, eager to catch up on work before the team arrived. Which is why, before 9 a.m. on a bright, startlingly clear late-summer morning, I was alone at my desk when I heard what sounded like a tremendous explosion.

Without really thinking, I immediately grabbed my purse and ran down four winding flights of stairs to the main WNYC administrative offices on the twenty-sixth floor. The door to the roof was open, and as I walked out I saw several colleagues staring openmouthed to the west, where the World Trade Center was on fire.

The image is seared into my memory, dark-red flames shooting out of the north side of the North Tower, and a shower of silvery fragments blowing into the air through the south face of the building. "I was sitting at my desk and I saw a plane—I think it was an American Airlines plane—and it ran right into the Trade Center," said one of the men standing on the roof. I grabbed him and we ran down to the on-air studios on twenty-five, so he could tell WNYC's morning news host, and our audience, what he'd seen.

The station's FM broadcast tower was on top of the World Trade Center, and the crash had knocked us off the air. But WNYC/AM was still transmitting, and everyone began to scramble to cover the story that was unfolding three blocks to the south of us. I ran to the newsroom to field calls from our reporters who were just coming in to work, and remember seeing my colleague Nuala McGovern staring out the window, saying "All those people, getting into work early on a beautiful Tuesday morning, and they'll never go home . . ." I think I'd been in shock; it was the first time I realized that there were people, just like me, who had been sitting at their desks when the plane hit.

Then the second plane crashed into the South Tower, making it clear this had not been a terrible accident. Some extremely brave producers and reporters remained, and Laura Walker, the president of WNYC, refused to evacuate the building, in order to keep the station on the air. I gathered a few colleagues and walked down twenty-five flights of stairs, emerging onto Centre Street to a view of both towers in flames.

On the way uptown with two colleagues, we heard a reporter on the radio say that one tower was collapsing. My friends turned around to see the structures fall. I stared straight ahead, not able to look, hoping that if I didn't turn around the buildings might somehow still be whole.

Most of the bridges and tunnels into and out of the city were closed; commuter trains were heavily delayed if they were allowed to leave the city at all. A colleague and I didn't reach our homes in New Jersey until seven o'clock that night. But we were lucky, we got home. Some of our neighbors did not.

The next morning, I was unable to reach my *Studio 360* team in New York, because the phone lines would not connect. But I was able to call Melinda Ward, my boss at Public Radio International, in Minneapolis, who was the person responsible for getting PRI to develop *Studio 360*. When we spoke that morning, I told her I thought we should just forget about the show, because what good can we do in the face of such awful events? Melinda said, "Julie, get yourself together. Of course you must produce a show—we need what artists can offer now more than ever!"

So I organized conference calls through Minneapolis, because no one in our area could reach anyone else in New York or New Jersey, and asked the extraordinary *360* producers to get to work. Because the Municipal Building and all of lower Manhattan were cordoned off, we couldn't retrieve any of our audio or equipment. But we found a studio in Midtown where we could record, and contacted the poet Marie Ponsot, who had written poems about war and loss, and the composer John Corigliano, whose first symphony was a memorial to friends who had died of AIDS.

A few days after 9/11, Ponsot and Corigliano were in our borrowed studio to talk with Kurt about art in the wake of terror. The conversation was wide-ranging and powerfully resonant. Melinda had been absolutely right; the program struck a strong emotional chord with our listeners, many of whom wrote to thank us for offering solace at such a terrible time.

It was also essential for my *360* team to put together that show. The work gave us a focus and a purpose in the chaotic days when we could not even get into the station; the music and poetry and conversation with these artists helped us and our listeners to begin to move forward again after the world shattered on that clear September day.

Artists can reweave the world for us when the fabric of daily life has been brutally torn apart. The following are three who have borne witness to some of the most devastating tragedies of the early twenty-first century and have worked to remember and transform terrifying events into transcendent music, theater, and photographs.

Terence Blanchard raced to evacuate his family from New Orleans before Hurricane Katrina hit, and returned with his mother to find her house in ruins. Blanchard later composed the eloquent and affecting music for Spike Lee's documentary about the aftermath of Katrina, *When the Levees Broke.*

Brooklyn-born playwright Lynn Nottage traveled to Uganda to meet women who had survived the relentless and cruel civil war in the Democratic Republic of Congo. Inspired by their stories of extraordinary tenacity in the face of rape and atrocity, Nottage wrote *Ruined*, her Pulitzer Prize–winning drama of pain and hope.

And Joel Meyerowitz spent months at Ground Zero, documenting the devastation of the World Trade Center as well as the extraordinary courage and perseverance of the workers who reclaimed the site. Each of these artists use their gifts to sift through the ruins, and to illuminate the resilience of the human spirit.

Trumpeter and composer
Terence Blanchard

Revealing a Private Tragedy

"The trumpet represents, in my mind, people on the rooftops crying for help and not being heard."

In one of the most devastating scenes in Spike Lee's account of Hurricane Katrina, *When the Levees Broke*, the camera follows Terence Blanchard and his mother, Wilhelmina, as they drive to see Blanchard's childhood home for the first time since the flood. Even before they open the wrought-iron door, Mrs. Blanchard begins to cry, and her son tries to comfort her. "You can rebuild this," he says, his voice breaking. "Oh, Lord," she says, "that's easier said than done." Inside, they survey familiar rooms made unrecognizable by the sodden destruction. After a few fraught minutes, during which they try to identify pieces of furniture that are upended and grieve over picture frames whose photographs are gone, Mrs. Blanchard leans her head on a doorframe and sobs. The anguish of the moment is heartbreaking. "When we came back here," Blanchard said, "that was one of the hardest things I've ever had to do, other than burying my dad."

Terence Blanchard is an only child, born and raised in New Orleans. "I grew up in Pontchartrain Park, in the Ninth Ward, but not in the lower Ninth Ward. I lived in the lower Ninth Ward when I was a little kid." He learned to play piano and trumpet, studying at the New Orleans Center for Creative Arts. Blanchard left the city to go to college, then joined Art Blakey's Jazz Messengers, replacing Wynton Marsalis as the band's trumpeter. Soon he was scoring movies for Spike Lee, a partnership that began in 1990 with *Mo' Better Blues*.

By the time Katrina hit, in 2005, Blanchard was back in his hometown with his wife and children. Before the storm, he hurried them to safety in Atlanta. "I tried to get my mom to leave," Blanchard said, "but you know with this city, man, we have this history of hurricanes turning on us and not hitting us directly. Katrina didn't hit us directly. I told my mom, 'You need to leave.' I said, 'Do you want to come with us?' She said no."

Like so many families, the Blanchards were separated by the storm and unable to reach each other for weeks. Blanchard finally found his mother in Mississippi, and convinced her to come live with his family in Los Angeles for six months while he wrote the score for Spike Lee's film *Inside Man*. In L.A. he got his mother her own apartment, and later said, "I was crying when I took her to Target and she was shopping for everything, like a college student. Everything—housewares, bedsheets, knives and forks, towels, clothes, everything. She had to buy it all. And I'm saying to myself, 'What the hell is this?' And the thing about it is that we were the lucky ones."

Shortly afterward, Spike Lee came to L.A. to work with Blanchard on the movie soundtrack and told the composer he wanted to make a documentary about New Orleans and its people in the wake of Hurricane Katrina. "I've always respected Spike, but when he did *When the Levees Broke*, my admiration went sky high," Blanchard said. "He's a person who is popular and famous and has a lot of weight attached to his name. He brought all of that to bear to let people tell their story. He gave the world a large glimpse into what had gone on here, from a lot of different viewpoints."

The four-part series is a brilliant, harrowing account of the storm and the ensuing chaos and disaster. Lee interviews residents who lived through the floods as well as public officials who wincingly attempt to explain what happened and why people had to wait days for rescue and relief. He includes chilling footage of survivors stranded in attics, or waving homemade flags for help as they stand on partially submerged bridges, as well as images of those who did not survive, floating through city streets flooded to the rooftops.

Lee wanted to include Blanchard's and his mother's viewpoints as well. "Spike asked her, 'Have you been in your house yet?'" Blanchard said. "And she said, 'No.' And he said, 'I want to film it.'" Mrs. Blanchard agreed, although her son was not so sure it was the right thing to do. "I was a little nervous about it. I talked to my mom about it. I asked her did she really understand what it entailed? She said yes, but she wanted to make sure that people would see what happened here. I was proud of her for doing that."

So in December 2005, Blanchard returned to New Orleans, driving with his mother and a camera crew to see what was left of his mother's house. "The only light that I had, going out to my mom's house, was the light from my car," Blanchard said. "To be one of the only cars on the highway—it's an indescribable feeling to think that the place you grew up in is basically a ghost town."

As he's done for many of Lee's movies, Blanchard wrote the score for the documentary, and his music eloquently emphasizes the emotions of disbelief and despair. Blanchard said that composing for this film was very rough going; at first, he felt he couldn't grapple with the enormity of what had happened.

"Everybody has been so heavily affected by this thing. It took us a while to even think about doing a project like this. People kept saying, 'Are you hearing music?' I said, 'No, man. It's so massive.' The only thing I could hear was silence."

Blanchard was finally able to move his family back to New Orleans in 2006, while he continued to work on the documentary. "On the one hand, I had to remain a professional and do my job," he said. "On the other hand, I was heavily affected by the footage I was looking at on a daily basis. Most people who watch the documentary watch it for about four hours, but I was watching it for weeks. When I took a break, I stepped out into the reality of what I was working on in the studio."

Early on in *When the Levees Broke*, there is a mournful dirge of strings with Blanchard's trumpet lamenting above them. Blanchard named the composition "Levees." "At the end of it," Blanchard said, "you can hear the strings, which represent the water, and the trumpet represents, in my mind, people on the rooftops crying for help and not being heard."

Blanchard also composed a piece titled "Funeral Dirge," music he described as his way of paying respect to the dead, mourning the deaths of people who might have been his neighbors. "When you see those dead bodies in neighborhoods that you know, that's chilling. Usually, when we see anything like that on film, it's in a war zone or somewhere across the ocean. Someplace other than here. But one of America's great cities, with that type of devastation and death, it's incomprehensible."

When the Levees Broke was broadcast on HBO on the one-year anniversary of the hurricane, and won a Peabody Award. A year later, Blanchard released a CD of music drawn from the film, as well as new compositions he and his band had written. "The name of the CD is *Tale of God's Will: Suite for Katrina*," Blanchard told us when the record came out in August of 2007. "It's all about New Orleans and Katrina. You can't avoid your daily experiences. You have to write about those. From an artistic point of view, I had to do it. It became a very emotional thing." The beautiful and heartfelt recording, full of pain and anger as well as glimmers of hope, won a Grammy Award.

"The thing that was interesting about this CD for me," Blanchard said, "was that I didn't want to do it, really. I didn't want to be a part of that whole movement of folks where when something happens, everybody tries to jump on the bandwagon. But at the same time, I started realizing that I am a part of the story. Being an artist, you can't avoid your social setting."

Blanchard ends *Tale of God's Will* with a tune he calls "Dear Mom." It begins with long held notes in the string section, violins and cellos playing a sweeping, sad melody over which Blanchard plays a trumpet solo of great power and delicacy. It's his tribute to Wilhelmina Blanchard, who again lives in the tidy brick house, now entirely rebuilt, that she and her son found ruined not so long ago. Many of her neighbors have not yet returned. "I was really proud of my mom." Blanchard said that his mother's willingness to allow Spike Lee to film their emotional return to her flooded house "was a hard thing to do . . . to take a very private moment and make it extremely public. ['Dear Mom'] was just a means to express musically what words just couldn't say to my mom."

Playwright Lynn Nottage

Listening to Untold Stories

"How does love continue to flourish
in the face of such ugliness?"

Tall, rough tree trunks ring the stage, surrounding a makeshift bar with mismatched, brightly colored chairs and rickety tables, at the beginning of Lynn Nottage's Pulitzer Prize–winning drama *Ruined.* In the sunlight, the trees appear to offer a little cover from the heat, but they turn threatening when raked with blue light in the dark. Similarly, the bar is a place of both shelter and danger for the women who work there. The imperious yet tender-hearted proprietor, Mama Nadi, demands that her patrons—the soldiers on all sides of the violent civil war and the miners who dig for the coltan ore that can make their fortunes—leave their ammo clips at the door. Mama's approach to the chaotic, ever-changing conflict that rages around her is to keep her eye focused on the cash box, which she believes will guarantee her independence. "If things are good," Mama Nadi tells her girls, "everyone gets a little. If things are bad, Mama eats first."

"Mama's Place is a bar in the Ituri Rainforest, which is in the

Democratic Republic of Congo," Lynn Nottage explained. "It's a place where miners and soldiers congregate to drink and release the tension of the war. It's also a place that provides refuge for women who have been traumatized by the war. Mama has provided them with a home."

It's a complicated refuge. Before they found their way to her doorstep, these women had been raped and sexually abused, like thousands of women and girls in their country. At Mama's Place, a chief's daughter named Josephine and a simple woman named Salima both work as prostitutes. The beautiful young Sophie, whom Mama almost turns away, has been so roughly treated by her rapists that she is mutilated, "ruined," leaving her unable ever to have sex or bear children. The play, Nottage says, "is the story of sexual exploitation of women, not just women who've been raped but what happened to them after they've been raped and the way they continue to be exploited."

The premise is gruesome and horrific, yet Nottage's characters often convey a buoyancy and optimism that belie the terror. When not working, they share stories, bicker, and dream of a future without war. "I wanted to look at the human side," Nottage said. "I was really interested in what happens to the human spirit. How does love continue to flourish in the face of such ugliness and brutality?"

Lynn Nottage grew up in Brooklyn, New York, and received her master's degree in playwriting at Yale. Then, instead of devoting herself to her art, she took a job with Amnesty International. "When I graduated I felt that I needed a much more expansive experience. I had spent the time from when I was in first grade through graduate school in an academic setting, and I felt I needed to interface with the world in order to write." Nottage was the national press officer for Amnesty, continuing a family tradition of activism. "My mother was on the front line. As a student she was in the civil rights movement and took a bus down in 1960 to be a freedom rider. She was always politically engaged."

While working at Amnesty, Nottage said she sometimes found it challenging to tell and retell the stories of people who were suffering around the world. "I feel you develop calluses when you're in a job like that. [When you deal] with human rights abuses on such a daily basis, and look at images [of abuse], you turn part of yourself off. Which is a horrible thing to say. But you have to in order to survive, because if you engage with every single image, you would not be able to get through the day. It doesn't mean that you're turned off to the horror. It just means in that moment, to get through it, you distance yourself from it."

Nottage knew she wanted to set her play in the Democratic Republic of Congo because "The conflict in the Congo seemed protracted and complicated. And [I was] frustrated that there wasn't an appropriate amount of attention being paid to that conflict by us, in America, who certainly benefit from the chaos going on there because of the resources that are being exploited. We use the coltan that comes out of that ground."

Coltan is shorthand for columbite-tantalite, a metallic ore found throughout the eastern part of the DRC. When heated, it becomes a superconductor. "It's used in cell phones, and it's used in computers, and it basically fuels new technologies." The mining—and the wealth it produces—exacerbate the bloody civil war that has raged in the country since 1998. It's estimated that more than five million civilians died during the first ten years of the conflict, from violence and disease. In October of 2009, Oxfam reported that during just ten months of fighting that year, "More than 1,000 civilians have been killed, 7,000 women and girls have been raped, and over 6,000 homes have been burned down in the eastern provinces of North and South Kivu. . . . In a bleak calculation by the [UN] coalition, for every rebel combatant disarmed during the operation, one civilian has been killed, seven women and girls have been raped, six houses burned and destroyed, and 900 people have been forced to flee their homes."

In order to develop a play that looked beyond these grim statistics, Nottage wanted to hear firsthand the stories of women who lived in the Democratic Republic of Congo. So she traveled to neighboring Uganda with the play's director, Kate Whoriskey, in 2004 and 2005. "We went to Uganda because there were so many Congolese refugee women pouring over the border," Nottage said. "We actually went in search of them. I was surprised by the number of women who wanted to share their stories. One by one they'd recount their narratives to us, and I began to hear some similar themes throughout. I think that in the process of listening to their stories, my play was being born."

After interviewing more than thirty women, Nottage found herself with hundreds of pages of transcripts. As she sat in her apartment in Brooklyn, reading through them, "I thought to myself, 'This play will be the ruin of me.' I knew I wanted to tell a story that was not agitprop, that was universal, epic, and unabashedly theatrical. Something truthful and yet joyful. And I didn't know how I was ever going to do that."

"It wasn't until I pushed myself away from that research and thought, 'Well, what does the story want to tell?' I don't want to do a verbatim play," she said. "I don't want to recount these narratives. You can pick up a Human Rights Watch report or an Amnesty report and you can hear what's going on. I really want people to know who these women are. Because when they sat down, most of the women said, 'We're telling you our story, because we never get to tell our story from beginning to end. No one knows who *we* are.' "

Black-and-white photographs of some of these women lined the walls of the lobby of the Manhattan Theatre Club, where

Ruined was performed in New York. Their eyes revealed awful pain, some filled with tears; others looked out with haunted stares, avoiding the camera. "One woman who really stood out was a chief's daughter," Nottage said. "Her story entered my heart and never left. She describes being in her home and being dragged out by a group of soldiers, taken to prison, being raped by four men, and using a watch on her wrist to bribe her way out, and then just running."

Nottage modeled the character of Josephine on this woman's experience, and named another character Salima after her. Midway through the play, Salima delivers a monologue in which she describes the morning that destroyed her quiet world of family, home, and work. She was tending her vegetable garden, her baby daughter playing under a tree nearby, just like any other day. "The sun was about to crest but I had to put in another hour before it got too hot," Salima tells Sophie as they get dressed for work in their tiny, shared bedroom behind the bar. "It was such a clear and open sky. How this splendid bird, a peacock had come into the garden to taunt me and was showing off its feathers! I stooped down and called to the bird, *ssst, ssst,* and I felt a shadow cut across my back. And when I stood, four men were there over me, smiling wicked schoolboy smiles. Yes? I said. And the tall soldier slammed the butt of his gun into my cheek, just like that. It was so quick, I didn't even know I'd fallen to the ground. Where did they come from? How could I not have heard them?"

"In that moment," Nottage said, "I was conjuring 9/11. That day, which was just such a clear, beautiful, absolutely perfect pristine day, and how suddenly our lives can be changed so absolutely in a moment." This wrenching experience, of having a normal day fill with disaster, is echoed again and again in the play. "A lot of [the women] spoke of not anticipating the moment, that suddenly the men were upon them. They were in the field or in their homes and someone kicked in the door and they were there," Nottage said. "One of the things that people don't know or understand is that there's an ebb and flow to the war. People are going about their daily lives, and all of a sudden soldiers will tumble into the town and chaos will break out, and [then] they'll retreat, and the people will be forced to pick back up the pieces of their lives and go about their business, not knowing when hell will descend again."

Why these women were willing to pour out their stories of atrocities and pain to a stranger is something Nottage has pondered a lot. "I think there was a sense of appreciation. Someone was willing to sit down and be patient and listen to their story from beginning to end. They felt like aid workers were not interested in their personal narratives. [The aid workers] wanted the quick highlights: 'What happened?' 'This happened.' 'Okay, here's your grain, move on.' Their families in many places had absolutely rejected them, and did not want to hear their stories. Many of the women said their friends and families weren't sympathetic. . . . Because it's a wound. They didn't want to see the maggot-filled, festering wound. Just cover it up, bandage it, we don't want to know that it's there."

The character of Sophie has physical wounds that won't heal, and she carries herself with the wincing, awkward gait of someone for whom every step is painful. Sophie is beautiful, and the men who frequent the bar often make advances, but Mama Nadi protects her. Sophie is smart; she keeps Mama's books, and sometimes keeps a little for herself. Sophie is also a singer of great emotion. Music fills this play, with propulsive drums and jangly guitars playing melodies that are often joyous. In a song that begins in an upbeat frenzy, and ends in a quiet plea, Sophie sings to the assorted violent customers: "Have another beer, my friend, wipe away the angry tears, my friend, get drunk and foolish, brush away the day's heavy judgment . . . and dance like it's the ending of the war." The song is full of affirmation and despair at the same moment.

Nottage wrote the lyrics, and she wanted music to play a powerful role in the story. "One of the things that I often hear," Nottage said, "[is that] people ask, 'How do you marry the beauty and ugliness that occur?' And I said I wanted to figure out a way to do that in the play; I thought the music and the songs were the most successful marriage in that regard." This complex collision appears within characters, too. Mama Nadi seems to be concerned only with herself, trying to make as much money as she can out of the turmoil. She has an acerbic, and sometimes hilarious, sense of humor. But she also offers a rarely glimpsed tenderness, particularly toward Sophie. The *New York Times* critic Ben Brantley wrote, "Ms. Nottage has endowed the frail-looking Sophie, as well as the formidable Mama, with a strength that transforms this tale of ruin into a clear-eyed celebration of endurance."

Although she spent just a short time on their beautiful continent, Lynn Nottage had no qualms about portraying their lives and creating the voices of these women. "I'm not African; I'm African American," she said. "When I was sitting with the women, I definitely felt a level of sisterhood, and I understood on some fundamental level that I was telling a universal story. Yes, this is specific to the Congo in this day and age, but I do believe at any moment it could happen in any place."

Given the subject, it's a triumph that *Ruined* is not a polemic. The characters work their way into our imaginations and our hearts, just as the playwright intended. "It was really important for me to have the audience engage with the piece. And I know the theater I enjoy and like. I hate being preached to. When I sit in the theater and I feel, 'Oh, I'm going to be lectured to, and here's your medicine and you're going to swallow it,' I turn off." Some who saw this play were astonished at the humor and joyousness that flash amid the gloom of war and atrocity. To those who question whether these elements can coexist, Nottage has an answer: "They have never been to Africa. They have never spent time with these women to understand that you can be brutalized and still find a way to heal. It was very important to me to be optimistic about that and still tell the truth."

Photographer Joel Meyerowitz

Documenting the Aftermath

"I don't want to see this history disappear."

For fifteen years, beginning in the early 1980s, Joel Meyerowitz had a studio on the twelfth floor of a building on West Nineteenth Street in Manhattan. The windows faced south, overlooking squat apartments and warehouses, toward the skyscrapers at the tip of the island, which were presided over by the two imposing towers of the World Trade Center.

Meyerowitz captured the view at all times of day in hundreds of photographs he took with an eight-by-ten view camera, and often returned to the studio even after he moved out in 1996 to continue shooting the Trade Center. One evocative image depicts the drama of the sky at dusk, as the thick bands of office lights on the Twin Towers grow bright against deep blue clouds.

On a calm, late summer day marked by a muted sky, Meyerowitz visited the studio to take another photograph of the Towers. "I made that picture, which was very quiet, nondramatic, and I remember thinking clearly to myself that it's not so interesting," he later said. "But I'll come back some other time. They're always going to be there." It was September 5, 2001.

Days later, the Twin Towers in smoking ruins, Meyerowitz approached the site of the destruction with his camera. "Like everybody else, I was curious to see what had happened down there. And like all the other passersby, I stood outside the chain-link fence on Chambers and Greenwich, and all I could see was the smoke and a little bit of rubble. I raised my camera to take a peek, just to see if there was something to see, and some cop, a lady cop, hit me on my shoulder and she said, 'Hey, no pictures.' And it was such a blow that it woke me up in the way that it was meant to, I guess.

"When I asked her, 'Why no pictures?' she said, 'It's a crime scene. No photographs allowed.' And I asked her, 'What would happen if I were a member of the press?' She told me, 'Oh? Look back there.' And back a block was the press corps, tied up in a little penned-in area, and I said, 'When do they go in?' And she said, 'Probably never.'

"As I walked away from that, I had this crystallization, probably from the blow, because it was an insult in a way. I thought, 'If there are no pictures, then there'll be no record.' We need a record. And I thought, 'I'm going to make that record. I'll find a way to get in because I don't want to see this history disappear.'"

Joel Meyerowitz is one of the preeminent photographers of his generation, with an extraordinary eye for color and light. He is also extraordinarily tenacious, a characteristic that served him well in the days after 9/11. Immediately after his encounter with the policewoman, Meyerowitz began working all his New York City contacts in order to penetrate the security surrounding the site. "I got the access by begging the director of the Museum of the City of New York to write a letter for me that would say I was making an archive. Then I got a friend who is now the commissioner of parks in New York to give me a red pass, which allowed me to go into the zone as a worker visitor. I hoped the letter would make people recognize I was making this historical archive, but in fact, most people down there didn't read or didn't want to read. I was stopped repeatedly, but I pressed on."

Meyerowitz is also a photography historian, and that role helped him figure out how to approach the awesome task of documenting the aftermath of the collapse of the World Trade Center. "I was prepared for it because over a ten- or twelve-year period, I worked on a history book called *Bystander: The History of Street Photography*, and that allowed me to be in archives all over the world. So I began to see what an archive looked like. The depth of it, the meaning, the necessity, the fullness, the description. And I thought, 'I know how to make an archive, because I've seen great historical archives—the Civil War, the Farm Security Administration of the Depression, Atget and Marville in Paris.' I understood something. So that focused my sense of being the right person."

Meyerowitz gained access to the site on September 23, and began spending most of his time there. "At least five days a week, minimum of four hours, up to ten hours. Sometimes it's hard to leave. Things come up. You see things that fulfill your idea of what should be in an historical record. The quality of life in the zone down there. A lot of people have come to work and have all kinds of services that they offer, and they should be recorded too, in the simplest way. So that should a future historian—or just a simply curious person—go to this archive in 2050 and look back fifty years, they could see the clothes that people wore, the quality of the light in the city, the look of the trucks and the hardware."

Over the next six months, Meyerowitz took more than 8,000 photographs. There are many images of the iconic, columned ruins of the first few stories of the South Tower that stood hovering above a pile of twisted, smoking metal, fire hoses spraying and backhoes digging through the charred and rusted remains. One photograph depicts the crumpled bronze globe statue commemorating the 1993 bombing of the North Tower, in the center of the rubble-strewn plaza, pierced by a beam, with the ruined façade as its backdrop. Yet another offers a panoramic view of the destruction at night, the ruins surrounded by smoke and haze.

Some of the photographs capture the enormity of the disaster, looking down on the sixteen-acre site while huge machines, tiny from this bird's-eye view, work to clear what seems like an endless pile of crumpled metal. Other pictures show the solemn, eerie emptiness of the streets near Ground Zero, where a thick white dust covered everything, from clothes in nearby shops to toy cars on the floor of an apartment.

What was it like, as an artist, for Meyerowitz to assume the role of a photojournalist in order to document this extraordinary event? "It creates tensions and it raises issues, photographic and aesthetic issues for me," he explained. "Do you come into a place like that with a preconceived aesthetic approach, or do you let that place tell you how the content should be described? Basically I found myself letting go of any of the trappings of ego one might have in making the pictures, and just letting the place tell me how to do it. Trying to make interesting photographs and descriptive photographs but also keeping an eye on those historically significant but maybe minor players. If you think for a second about any historical photograph you might have seen, say, of 1900, a picture of Twenty-third Street and Fifth Avenue, there are trolley cars, and horse-drawn wagons and strange lampposts. The fire hydrants are different. The cops are wearing little belts and tunics with skirts on them; the signage is different. All that stuff is background, and even in the World Trade Center, in the background of the things there is content. In a way, as an historian and an artist, I had to try to deepen the field of my observation and my interest, which meant sometimes letting go of a big dramatic image up close in order to let the background speak as well as the foreground. I found myself opening up and maybe even reinventing a way of looking for myself. Maybe this is a really deep learning experience. It is, for sure. Maybe aesthetically I'm evolving again."

In many of the photographs, Meyerowitz pays close attention to the people who work here, taking pictures of them as they attach winches to gigantic fallen I-beams, tend to the huge rigs that lift the metal remains, stand in tan overalls, a sledgehammer on one shoulder, near a giant earth mover, or carefully rake through what is left on the ground of the giant hole for any possible human remains. One dark image shows the site at twilight, with just the shadows of people in the foreground. "It was just after dusk, the sky had gone from pink to plum to purple, and I was standing around with a bunch of guys from the arson explosion squad, and they were my angels. They protected me. Basically without them I couldn't have done this work continuously. We were standing on a little hill that they made in the middle of the pile, because they had to make their roads to get in. And way down in the bottom, coming from the bottom we hear the plaintive sound of taps being played on a bugle. Down there was a guy with an American flag tied to his knapsack, and he was blowing this lament for the dead. All of us were teary and had goose bumps. It was chilling. It was the kind of thing that if we saw it in a movie theater, in that kind of enormous Colosseum-like space, we would be weeping. The whole theater would be weeping."

Work on the pile was rapid, and the changes dramatic. Cleanup had been expected to take a year; instead the site was cleared in about nine months. Meyerowitz said that his process changed as the site changed. "In the beginning, I really wanted five other photographers to be down there with me because I thought it required a task force to do this. There were so many things that were being taken away so quickly. But the city wouldn't even allow me in, to allow five was luxury beyond expectations. So I found myself racing to get to things before they took them away—the trees, cleaning up the cemeteries [behind St. Paul's Chapel, where George Washington once worshipped, across the street from the World Trade Center site], clearing out the rubble in the offices, the side streets. I was trying to record everything. I found myself racing from point to point, and the pile itself was such a spectacle, I would make my perimeter journey every day, maybe twice a day, around the pile, and across it where I could. But slowly—actually, not so slowly—fast, they've removed tons of rubble, and the pile is now a hole, a deep hole, so the look is different, what's to see has changed. It's no longer a drama of this big spectacle; it's how they excavate to get down to train platforms, and the tubes, and parking lots. It's a different, more mysterious place now, and they've reached the point where the 'bathtub,' which is the enclosure for the World Trade Center foundation, which held out the Hudson River, that is the thing that's been revealed to the point where they're now shoring it up—they have been for months. With long pipes and forced concrete, so that this wall doesn't give in and let seepage in. That itself is like some enormous Mayan ball court, or it feels like the Pyramids. There's something profoundly ritualistic about it."

Meyerowitz has been deeply changed by the experience of photographing this place, day after day. "First, I think that I feel again at my age now [in his early sixties] the need to be more useful as an artist. There was a period in my younger days when I was socially concerned and committed and did things about the Vietnam War and about our life; and over my middle period, you might say, I could feel my connection to the medium of photography and the questions I was asking about why photography looks this way, how it works, how it describes things. Sort of art for art's sake. And over the last ten years I've been feeling that need to do something more utilitarian and socially conscious. So I feel by rising up and saying I'll do this without funding, I haven't gotten any funding yet, neither has the museum, I'm dangling in the breeze basically, but I couldn't wait around for the proposals to be written, I needed to do this at whatever cost.

"So, on the one hand, I feel politicized. On the other hand, there's a spiritualized quality to that zone that's undeniable. You enter it through all the air locks, all of the chained-off gates, you finally get in, and every day you pass the same pile, at least in the beginning, and standing there and realizing that there's a thousand or more dead people chewed up in these remains. And it's awesome. And deeply moving. And everything one does in there—I'm talking about the workers, the crane operators, the riggers, the burners, the policemen and firemen—everyone has gone about this with a kind of tenderheartedness that's so apparent that it connects everybody. There's a kind of religious quality that's not defined by any particular religion but just by the sense of spirituality. I feel malleable, softer in some way, more respectful."

After one of his photographs appeared in *The New Yorker* magazine in October 2001, Meyerowitz was contacted by a representative from the U.S. State Department, which was interested in creating an exhibition of his photographs. "They said, 'We want to send a show around the world to our friends and our enemies'—that's exactly how they said it. They want this work to go to Muslim countries. Everywhere. So they could see that it really happened. That this was the result of this terrorist attack, that it wasn't Hollywood or a concoction. And that we're healing ourselves, we're going about the business of recovery and salvation on many levels." Six months after the attacks, the State Department opened exhibitions in more than twenty countries.

Meyerowitz published many of these photographs in 2006 in a book titled *Aftermath*, and a number can be seen on his website.

His pictures are viscerally powerful, with such a punch that it was years before I was able to look at them without weeping. They bring back for me, and, I suspect, for anyone who spent time near the site in the months after the attacks, the awful loss I felt, the acrid smell that permeated my clothes and hair after a day at the office, the wrenching image of ruin, frenzied work, and, finally, emptiness that hit each time I walked past Ground Zero on the way to the radio station. Yet some of the photographs are also profoundly, unexpectedly beautiful, capturing the astonishing light as well as the wreckage. In his conversation with Meyerowitz, Kurt asked, "When you produce some of these images, which are so astoundingly beautiful as pictures, is there a problem given the subject matter? To show something so ugly in such a gorgeous fashion?"

Meyerowitz replied, "It was ugly—powerful and tragic and horrific. But it was also, as in nature, an enormous event that was transformed after the fact into this residue. Like many other ruins, you go to the ruins of the Colosseum or the ruins of a cathedral someplace, and they take on a new meaning when you watch the weather. There were afternoons I was down there, and the light goes pink, and there's a mist in the air, and you're standing in the rubble, and I found myself recognizing both the inherent beauty of nature and the fact that nature as time is erasing this wound. It will smooth itself out. . . . Time is unstoppable. And it transforms the event. It gets further and further away from the day, and light and seasons temper it in some way. It's not that I'm a romantic. I'm a realist. The reality is, there's the Woolworth Building, in a veil of smoke from the site, but it's now like a scrim across a theater, and it's turning pink! And down below there are hoses spraying, and the lights have come on for the evening, and the water is turning acid green because the sodium lamps are on. And I'm thinking, 'My God, who could dream this up?' But the fact is, I'm there, it looks like that, you have to take a picture."

In *Studio 360* the week after the 9/11 attacks, Kurt asked the poet Marie Ponsot what artists can do when faced with such destruction. She answered immediately, "I think it's a very sharp moment for people in the arts, those who love the arts, those who make them. I think it asks particularly of people who make art a very poignant question: If you think that art is not worth doing in a time like this, it probably isn't worth doing at any time. If you think that art is indeed part of what I call the world's work, then to be loyal to it and to look to it for strength, for its strength now, seems right."

Chapter 9

GETTING TO WORK

The summer after my freshman year in college, I worked on an archaeological dig. But I almost didn't make it to the site; while the plane was boarding, my parents and I frantically searched JFK for a mailbox so that I could send in my last term paper, which I had finished writing in the car on the way to the airport. The paper was three weeks late. It had to be postmarked that day or I would fail the class.

During high school and college, I was an incorrigible procrastinator. People seemed to have high expectations for me, but I doubted I would ever achieve what they thought I was capable of, and procrastination gave me an excellent excuse for never living up to these standards. "Oh," I would think to myself, "I began this twenty-page paper last night at eleven, and wrote through the night, so if it isn't very good, well, I just didn't have enough time."

My junior year in college, I discovered two secret weapons against procrastination in exactly the places I frequented to avoid schoolwork: the ceramics studio and the radio station. I spent most of my time playing with clay during my last two years of school, preparing for my senior show and reveling in the exquisite, tactile respite from writing papers and taking exams. I also learned an essential lesson: if I made a mug or vase that I didn't like or that didn't quite work, all I had to do was squish it back up into a ball of clay and start again. I loved to make things, and to destroy them, knowing I could try again.

This was a revelation to me. I had always thought that the first thing I made or said or wrote down had to be perfect (maybe because I had never left myself time to think of a second thing). I began to realize I could take the same approach I had with clay to my other work, perhaps even revising a paper before handing it in.

My experience with clay helped disarm my desire to make perfect things, but it couldn't entirely cure me of procrastinating. I needed a deadline I couldn't push—and found that in radio.

At my college radio station, I worked a weekly shift as a DJ. One night late in my senior year, as I played medieval chants, twentieth-century cello concertos, and Grateful Dead songs back to back, a friend called me on the phone in the studio. "I was listening to the radio," he said, "and heard this woman with a beautiful voice. And then she said she was you! You should do this for a living."

A short time later I spied an ad in the career resource library notebook for a production assistant job at WNYC. Two days after graduation, I began work at WNYC on a show about art and culture in New York. I loved immersing myself in the cultural life of the city, interviewing artists, musicians, and writers about their work. The job also broke my procrastination habit. Fast. In broadcasting you can't ask your listeners for an extension.

It was the era of audiotape, and one of the first things I learned in crafting radio stories and programs was how to back-time—how to make a piece of music end at exactly fifty-nine minutes past the hour, or get a particular sound effect to hit at exactly the right moment by figuring out how many seconds you have before you want to hear it in the clear. From this I learned how to backtime my work, how to organize all of the effort so that I was ready to go on the air when the clock hit my mark.

I chose to work in radio because I love talking to creative people about their work, and love opening up the world for listeners through conversation and sound and music. But I see now that radio also demanded that I change the way I work, and gave me a framework to learn how to do that. I was fortunate to find a field that wouldn't let me ask for just a couple more days.

In *Studio 360*, many artists have described for us key elements they need in order to work, among them Yo-Yo Ma, who describes finding something new in the familiar. Isabel Allende reveals her secret to beginning a book. And Tony Kushner talks about wrestling with a play until it's finished—or abandoned. Rather than provide "how to" tips, these artists offer insights into how they get to work, how they deal with the inevitable moments when work is going badly, and how they know when their work is complete. Their stories resonate for the rest of us, regardless of the creative passion we choose to pursue.

Cellist Yo-Yo Ma

Warming Up

"I become friends with the instrument."

My mother often reminds me that beginning a new project doesn't start when you sit down at your desk to write, or stand in front of a canvas with a palette full of paint, or figure out a new tune on a piano. For most creative work, there's a period that she likes to call "pawing the earth," when we must create the environment in which we can begin. Like puppies walking round and round the spot they've chosen to lie down in. Or Thoroughbred horses, held at the gate before the race begins.

For musicians, this work happens on a daily basis, when they sit down to practice. Kurt talked with Yo-Yo Ma about his routine, whether he's getting ready to perform a classical concerto with one of the great orchestras of the world, or preparing for a recording session of music from China to Iran to Uzbekistan with musicians from countries along the Silk Road. Wherever he is, Ma said, he begins each day reacquainting himself with his beautiful Montagnana cello, made in Venice in 1733, which he's nicknamed Petunia.

He starts by playing long, slow notes, which he demonstrated. "I become friends with the instrument," Ma said. "I try not to tax it too much. It's really like warming up a car, or warming up your body. You stretch it. You don't go into a fast run. You don't take it to sixty in three seconds. Because what's funny about an instrument made out of wood, every day the humidity is different. Every day the temperature is different. And wood—as well as our bodies—is slightly different."

By starting slowly, gauging how his cello feels that morning, and how he does, too, Ma reestablishes his relationship with Petunia. Before he addresses whatever he needs to rehearse that day, he plays a Bach cello suite, "something that I started learning as soon as I started playing. It's also something written for cello alone. This music is somewhat meditative. I think of the flow of water. Light, afternoon light playing on leaves, so that you see something that is familiar and yet different every day."

Kurt seemed a bit skeptical, and wondered whether a piece that Ma has played hundreds and hundreds of times could really reveal something each time he plays it. "What's amazing," Ma responded, "is that with a great friend, you could see them thousands of times. And you don't look at them one day and say, 'Well, today, I'm really bored with you.'"

Ma was four years old when he began to play Bach's cello suites. "When I was a teenager, I looked at the score, and it's written in different bowings, so I tried that. And then at some point the head of the Schweitzer Institute wanted me to talk about Bach. Suddenly, having to verbalize music that I'd known forever took me to another way of thinking about it. Bach, as Schweitzer once described, was a pictorial composer. And then I thought, 'Well, what does that mean?' One of the things [Bach] coded was infinite variety. Instead of the materiality of saying,

'This is the same thing, I need a new product,' it's something new every time." Yo-Yo Ma makes clear the power of his connection with both his instrument and his music; from them, in his warm-up, he draws the intensity and depth his performance is famous for.

Novelist Isabel Allende

Creating a Ritual

"My life is complicated, and if I didn't have a time to start, I would never do it."

For other artists, the work before the work is often research, or finding a quiet place in which to create without interruption. Once that's accomplished, the next challenging step is actually getting started. This can be a fraught piece of the process, one that invites ritual to help order the experience of beginning something new.

The writer Isabel Allende has a very clear ritual. When she visited *Studio 360* to talk about her novel *Inés of My Soul*, she said that every year, on January 8, she sits down at her desk to start a new book. "I haven't finished the one that I started this year, but I'm going to start another book all the same, because if I don't start on January eighth, I would probably not write it. I want to start something and then finish what I'm doing, then go back to the book I will start next year. But I'm still doubting, doubting what that book will be about. Can you imagine what January seventh is like in my life?"

Each year on January 7, Allende organizes the desk in her study beside the pool at her glamorous house in northern California, clearing away all the materials that relate to her old books, leaving only the research that she's conducted for a new novel or memoir, although sometimes she doesn't yet know what she will begin. On the morning of January 8, she meditates and lights a candle before sitting down at her computer to write. "I'm just procrastinating," she said. "It's so hard to just sit down and begin something." Allende waits for the first sentence to come to her, trying not to think about it. "The best writing comes like an instinct," she explained. "The more you handle it, the less fresh it is. That first sentence gives me the tone for what I'm going to write."

Allende spent four years researching the life of Inés Suárez, and when she began writing one January 8th, she "sat down and turned on my computer and the voice of Inés came to me immediately. I didn't have to think how I was going to narrate the story. It was in first person. Her voice. The book starts when Inés is already over seventy and she knows she is going to die soon. She's remembering." It's a strong, unexpected voice for a woman from sixteenth-century Spain. "Few have heard of her because history is written by males. The ones who win the battles. And all the people who are defeated, especially women, are left out. Inés Suárez was the only Spanish woman who accompanied 110 Spanish soldiers to the conquest of Chile. They had to cross the driest desert of the world and encountered innumerable perils. The lives of the 110 men are well recorded. Very little is said about her, although the few things that we find between the lines are astounding. This woman was a dowser; she could find water in the desert and saved the troops. She saved the city of Santiago from the first major attack of the Indians. She lived to be seventy-three years old, which was a very long life. And stayed in Chile and became the second wealthiest person in the country."

Allende said it was hard to imagine what the world that Inés

Suárez knew five hundred years ago was like. "When none of the modern commodities we have existed. Everything was rough; there was a lot of suffering. People lived short lives. And people were extremely brave. The war was not like war today, with technology and killing each other and in the process killing more civilians than soldiers. At that time, it was face-to-face. The death and wounds and confrontation were brutal."

Allende began her career as a journalist in Chile. In 1973, she fled the country after the coup that overthrew the government of her uncle, Salvador Allende. For thirteen years, she lived in exile in Venezuela, and her commitment to beginning new work on January 8 began there, in 1981, when she chose that day to start a letter to her grandfather. He was still in Chile, nearly one hundred years old and very ill. Allende intended the letter to be a farewell to a man who had been a large influence on her life and imagination.

Augustín Llona Cuevas soon died, but Allende found she couldn't stop writing the letter. As she recalled in her 1995 memoir, *Paula*, "Other voices were speaking through me; I was writing in a trance, with the sensation of unwinding a ball of yarn, driven by the same urgency I feel as I write now. At the end of a year, the pages had grown to five hundred, filling a canvas bag, and I realized that this was no longer a letter." Isabel Allende found that she had just written her first novel, *The House of the Spirits*. Since that time, she's begun each memoir, novel, or collection of stories on January 8. "It became a good luck thing, and then later it became a matter of discipline. My life is complicated, and if I didn't have a time to start, I would never do it.

"I became a writer out of necessity, and I did not quit my day job until I had three books published in several languages. I think the fact that I started late determined the way I am as a writer. The discipline, the perseverance. I want to make up for the time that I lost." Once she begins to write, Allende said, "I live like a monk. I get up in the morning, walk the dog, and lock myself in my room for the rest of the day." In her devotion to her grandfather who inspired her first novel, perhaps one can see a foreshadowing of the devotion to work itself that is characteristic of Allende's creative life.

Novelist Joyce Carol Oates

Thinking and Dreaming

"I don't write anything until it's ready to be written."

Joyce Carol Oates embodies a similar devotion to her work. She has confessed that she's often at her desk for twelve hours or more a day when writing a first draft. But Oates doesn't have a particular ritual to begin something new; if she did, she would have to perform that ritual quite frequently (she's often published two or three books a year and has written more than fifty novels and thirty collections of short stories, as well as volumes of poems and plays). Instead, she says, "I don't write anything until it's ready to be written. I do a lot of thinking. And until I have the whole thing in my head like a movie, I don't really begin it."

In her novel *The Gravedigger's Daughter*, she tells the story of Rebecca Schwart, daughter of refugees from Nazi Germany who settle in upstate New York, "a fictitious character who was generated by my memories of my grandmother, my father's mother, and the experiences that she had, really a long time ago. Before I was born. Most of my writing is generated by my memory or some episode out of history, a personal history or other people's experiences."

Oates said that immersing herself in these experiences, and dreaming about the story she will write, can be very emotional. "It's an intense and ineffable experience which is difficult to talk about. I think it's understandable in terms of art—you can create music that's very powerful and engrossing, and all-consuming, but you can't really talk about the music, you have to experience it."

Kurt then asked, "So when you sit down in the morning, it's like going to a job—and over time as you write it becomes an emotional state?" To which Oates replied, "Well, I don't think it's like a job. Which job are you thinking of? Basically, if you're an artist you've done a lot of thinking. You do the work in your head. Your imagination is basically something that you enter into like a dream. So when you go to take down a dictation from your memory—I write in longhand—it's more that you're remembering something that you've already worked on. I don't really make something up as I'm typing. I've never done that. It's more like memory."

Her key challenge at this stage, Oates said, is to put the dream world into sentences in order to communicate what she's constructed in her imagination. "It has to have a component that's very different from a dream. Dreams don't have any language. So you have to create in words a verbal structure that expresses, that has a certain resonance. . . . That's the hardest part. We can all have daydreams and we all have memories. But to transcribe it in a formal structure—I find that very challenging. And that's the level at which one might feel some anxiety."

In an essay for the *New York Times*, Joyce Carol Oates revealed her secret to combating that anxiety. "The structural problems I set for myself in writing, in a long, snarled, frustrating and sometimes despairing morning of work, for instance, I can usually unsnarl by running in the afternoon." Oates wrote about the many other writers who turned to running, or hiking, or walking, to open up their imaginations, like Wordsworth, who took long walks along a lake, or Thoreau, who "traveled much in Concord." Running for her is like the dreaming she does as she conjures up her stories. "Dreams may be temporary flights into madness that, by some law of neurophysiology unclear to us, keep us from actual madness. So, too, the twin activities of running and writing keep the writer reasonably sane and with the hope, however illusory and temporary, of control." Oates's dedication to writing, and to running, exhibits her characteristic discipline.

Sculptor Richard Serra

Taking a Break

"We had a four-o'clock problem."

Many artists who have visited *Studio 360* have spoken about the difficulties inherent in making things up. As Joyce Carol Oates suggests, an antidote to these moments, when creativity hits a wall, when an inner critic can derail the best intentions, is to move away from your work. Sculptor Richard Serra described what helped him combat what he calls the "four-o'clock problem" that he encountered in his early days making art in a downtown loft in New York City in the 1960s.

"Phil Glass was my assistant; he was working as a plumber. We were both moving furniture at the time," Serra said. "We used to just talk about what the nature of a problem was and what the nature of matter was and what the nature of time was and what the nature of process was. Then we would go through all these experiments, and at the end of the day—strange, I haven't thought about this in a long time—we'd get on the ferry or we'd get on the subway. Because we found that if we took ourselves out of the studio and got into a space where you didn't have to walk and you were being transported, that actually ideas were exchanged more rapidly. So we used to take ferryboat rides; sounds strange but we did. Or we used to get on the subway, and ride back and forth, then go back to the studio, just to get ourselves in a different mind-set. I don't think I've ever told that before. That was very useful.

"If you're right in front of a problem every day and you're not getting anywhere with it—you get backed up and you're getting constipated—you think, 'Hey, I'll just blow it off and get on the ferry and we'll talk about it in another way.' Or we'll get out of our assumptions about what is what and our predetermined philosophy that has been handed down from someone else, and try to open this thing up. We didn't consciously say that to each other, and we didn't consciously say, 'We're gonna think in a different way on a ferryboat ride.' But as it turned out, we used to say we had a four-o'clock problem. At the end of the day between three and four we'd always think, 'Okay, we're going to get it now.' Then we'd take the ferryboat ride right around then. We had an idea after a point that we were into a kind of symbiotic flow in terms of how we were doing with each other's head, and what an invention entailed." What might look to some like evasion becomes a liberating change of perspective that allowed Serra to see his work in a new light, and to move forward.

Novelist John Irving

Doing It Over

"When I first said to my wife, 'I'm going to take this book back from Random House,' I think she may have looked at me as if I had an undetected stroke in the middle of the night."

For some, the creative challenge may hit every day at the same time; for others, the crisis may arise at the same point in the creative process for each work. Sometimes, however, it's not possible to recognize a problem until the work is complete, when you've had the opportunity to get some distance from the piece. That's the scenario John Irving described to us regarding his novel *Until I Find You.*

The book is huge, more than 800 pages long, and tells the story of a man named Jack Burns, who traveled through Europe with his mother when he was a small child, in search of his missing father. They abandon the search, and Burns grows up in Canada and the United States. Irving portrays Burns's childhood, his yearning for his lost parent, his life as a successful actor, and how after his mother's death he picks up the search for his father,

"Who of course is there waiting for you in the final chapters. If he weren't, what a lousy story it would have been, if you didn't finally get to find him," Irving said. "I know the end of every novel before I start. That leads me to where I want the reader to begin the story. In this case, chronologically, but that's not always the case. In this case, because of how long and complicated I knew it was, I thought I better do at least one easy thing for the reader, and make it a linear story, which it is."

Initially, Irving had envisioned that the main characters were Jack's parents, "because so much of the beginning of the novel is about Alice, the mother, and so much of the end is about William, the father. What I'd overlooked was how important the middle of the story was. In the middle of the story, it was Jack's story."

Irving had written the book in the first person, from Jack's perspective, and it was only after he had already sent it to his editor that he came to the conclusion that this approach would not work. "I realized at the end of this long work that the book is more about Jack Burns than Alice or William. As compelling as characters in a novel as Alice and William are, the main guy is Jack. I felt I could bring more sympathy to Jack, more tolerance, more understanding to Jack, if I were in the third person voice and I got some distance from him," Irving said. "I would never write a novel in the first person if the first person narrator were the main character. The first person narrative suggests that the narrator is telling a story about someone who, in his mind, is more important than he is."

Irving decided to rewrite the entire novel. His decision met with initial skepticism from his agent, Janet Turnbull—who also happens to be his wife. "When I first said to my wife, 'I'm going to take this book back from Random House,' I think she may have looked at me as if I had an undetected stroke in the middle of the night. But I didn't really need to work very hard to convince anybody."

Irving said that this has never happened to him before. “It was such an enormous undertaking to rewrite this book from the first person to a third-person novel. Took me nine months, all by hand. I hope I come to that conclusion sooner if it happens again.” With *Until I Find You*, Irving was able to approach his work with a clear eye, recognize a problem, and have the courage to rework it.

Playwright Tony Kushner

Knowing When to Stop (and When to Keep Going)

"It was going to be one of those plays
that you could just spend the rest of your life
desperately trying to get right."

If a book is ever to reach its audience, revisions must at some point come to an end, a manuscript must be sent to a publisher, and the book printed and bound. But a playwright can tinker forever, and that nearly happened to Tony Kushner with his play *Homebody/Kabul,* which opened in New York in 2001, shortly after 9/11. Over the following three years the playwright revised it numerous times before its next New York performance in 2004.

Tony Kushner came into *Studio 360* that year, after *Homebody/Kabul* had a run at the Brooklyn Academy of Music, and when his musical *Caroline, or Change* had just premiered on Broadway and was touring the country. He was our guest for a show that asked, "When is it done?" He announced, "I am done with both *Caroline* and *Homebody/Kabul.* I'm done, I mean *Caroline* is actually finished, and *Homebody/Kabul* is abandoned."

"Abandoned?" Kurt asked. "That sounds negative and desperate." Kushner replied, "It doesn't feel desperate, and it's not entirely negative, although I wish I had another five or six weeks to work on it. Maybe I don't mean abandoned, maybe I mean 'finished for now.' I promised myself, since I've been working on this for seven years, when the play opened at BAM, I would stop. And I knew when I first started writing it that I would have a lot of trouble stopping. So I know it's really time to honor that promise."

Is it in the nature of this kind of work to want to keep going and going? Kushner said he thought that might be true. "I could sort of tell at the beginning that I was launching into something that I couldn't see the end of, that I was trying to pull together disparate ideas, story lines, thematic elements. I could feel that writing the play was going to be very hard, a huge gamble. I could feel it was going to be one of those plays that you could spend the rest of your life trying to get right.

"Every time I've watched the play—this doesn't happen with every play, and it's sort of scary when it happens—every time I've watched *Homebody/Kabul* I think, 'Oh, if I just put this old line that I took out in here, if I move this plot point around, I can make the moment, the scene, the whole play clearer.' Not the whole play, because the hour-long monologue that starts the play is as finished a piece of writing as any I've ever done. But the two hours and forty minutes that come after the monologue—even though I've promised I won't work on the play any more, I already have a set of rewrites in my head, and if I allowed myself I'd be tinkering again. I could do a little set of rewrites at least for the next production. But it's time to stop."

While he was working on this latest version of *Homebody/Kabul*, Kushner, with his collaborator, the composer Jeanine Tesori, completed work on their musical, *Caroline, or Change*, which the *New York Times* critic Frank Rich said was, like Kushner's most famous work, *Angels in America*, "about that sensation of history cracking open." In the musical Kushner tells a very particular story of a young white boy growing up in Louisiana in the 1960s, and the black woman, Caroline, who works for his family after his mother dies of cancer. Through these characters, Kushner probes a world in which change is coming, but no one knows what it will mean.

Near the end of the play, Caroline sings a song called "Lot's Wife," which begins mournfully with just voice and piano, and builds to a loud full band with horns and an organ, dying away again to just Caroline and a piano as she pleads with God to turn her into a pillar of salt, to end her desires for happiness and love and, as she sees it, set her free from the rage and discontent that are threatening her family and her life. The song is a repudiation of change, of transformation that brings progress. "That song is what musical theater people call the eleven-o'clock number—it's this huge moment toward the end of the second act, before the finale, usually when the main character arrives at a turning point, a reconciliation with herself. The musical has the word 'change' in the title, and none of us who was working on it wanted to make a song about despair, or surrender. But at the same time, in theater we're conditioned to expect a play or musical to state at the beginning that Character X has this or that problem, and an hour and a half later Character X has figured out what her problem is and is on the way to better things.

"Of course, it's weird, this expectation, because no one changes that quickly. No one changes irrevocably, with any absolute certainty. There's always the possibility of backsliding. And most people spend their lives struggling mightily to change a tiny, little habit like chewing their nails. So why do people who go to the theater, read books, go to the movies, expect to see people revolutionize themselves within a very short time? I think it's false entertainment. It's anodyne—promising things that don't happen. I think it only produces intense, private, personal despair in the audience because one knows one is incapable of such transformation. [Caroline] is a character who struggles mightily but can't or doesn't transform.

"I felt I had to insist on that as being the key to 'Lot's Wife,' but I was worried about my conviction, and there were endless discussions involving everyone, and the form of the song changed over and over again as what we were trying to express kept eluding us. There are seventeen versions of the song, a few that are happy and optimistic. One sounds like a graduation speech.

"We were still working on the song up until previews. It was the very last thing still being completely rewritten. One afternoon there was this huge group meltdown, everyone frantically suggesting ideas about what the song should say, and I took notes, and went home. On the subway platform I called George [C. Wolfe, who directed the play], and he said, 'I want you to go home and forget everything that everyone else said. Everything that I've said. Do your own version of this. You're the only one who's going to really nail it for the words, and bring it in for Jeanine, and we're just going to drop out now.'

"I did what George told me to do. That night I wrote the lyrics that turned out to be the finished version, and I brought it to Jeanine and I think literally within an hour it was set to music; she played it through for just her and me, and we both knew immediately, 'That's it.' We brought George in, he listened, and he agreed. Tonya [Pinkins, who played the lead character] learned it in two seconds—she's a very fast study. And that night, the audience, when she finished, screamed. And we thought, 'Oh, that's done. It works.' When it works, it works. It felt really good. I felt like a real Tin Pan Alley songwriter.

"After all that struggling we finally came back to a version of the song close to the original idea for it, but in a radically different form, and even the idea's much more developed and nuanced—for which I think we're indebted to all seventeen versions of the song we'd made. The struggle to write it is as great a source of pride for us as the effectiveness of the song."

In Tony Kushner's story, one hears both the painful uncertainties of the creative process, and the triumph of having worked through and learned from them. He received the crucial gift of being told, "Don't listen to anyone else—you're the only one who can do this."

Painter Chuck Close

Celebrate!

"I have traditionally, for thirty-five years, celebrated the end of every painting."

What, then, should be done once a project is complete? Chuck Close has a perfect answer: celebrate! "Every time I finish a painting, I play Aretha Franklin full blast, from start to finish, and I usually sing along with her as my celebratory end to each painting."

We often have so much else left to do once we have finally completed our work. But Close reminds us that we should never miss this essential step of celebrating the accomplishment.

There are as many approaches to the process of creating as there are creators. The artists in these pages offer ideas, and a bit of reassurance: many of them find it hard to get started, and sometimes despair of ever finishing what they do begin. But these are other people's stories; the work now is for each of us to create our own.

Acknowledgments

Writing a book is a rather solitary pursuit, yet this one emerged from the highly collaborative realm of radio, and so there are many people, in overlapping circles, who have been indispensable in helping me bring *Spark* to life.

I'll begin with thanks to the dedicated people who worked with me as I translated these *Studio 360* stories from radio to the page: my assistant Keri Wachter, whose research skills, new ideas and positive energy contributed to every step of the process; my "book doula" Laureen Rowland, who offered guidance as I figured out each chapter and edited many drafts, and was a wonderful mentor as I learned to write for the eye as well as the ear; my editor, Julia Cheiffetz, who offered excellent big picture direction and early, on encouraged me to tell my own story as a bridge to the artists on these pages; Katie Salisbury, who made sure all my questions found answers and patiently shepherded the book through each phase; Michael Morrison for meeting with me when things were up in the air and for sharing a chapter with his son; Kathy Schneider, Tina Andreadis, Samantha Choy, Nicole Vines, Nicole Reardon and their colleagues for getting the word out about *Spark*; Tom McNellis for shepherding the manuscript through production; Ed Cohen for his keen eye and sense of humor; Joy O'Meara for her lovely design; Mark Ferguson for his Web savvy; Bob Miller, who decided that the stories in *Studio 360* would make a great book during a conversation over dinner with Anne Kreamer, and Anne for encouraging that idea;

Heidi Schultz for her good-natured persistence in tracking down permissions; Ken Norwick for his wisdom and support; Stuart Krichevsky for his wise and patient guidance and firm advocacy for the book; and the Maplewood Library for endless interlibrary loans.

The next circle includes those to whom I owe so much for their work on *Studio 360* over the past ten years. My thanks, first, to Melinda Ward at Public Radio International, who came up with the idea to create a program about art and culture for public radio and has championed the program, and now the book; and Laura Walker at WNYC, who offered constant support to me and to *Studio 360* and has been an extraordinary colleague and dear friend since we both began our lives in radio.

My great thanks to *Studio 360* host, Kurt Andersen, who has given voice to the program, conducted most of the interviews I've drawn from in this book, and was a wonderful partner in developing the ideas for *Spark*. I am tremendously grateful for the creativity of the talented *Studio 360* team, past and present: senior producers Kerrie Hillman and Leital Molad; senior editors Peter Clowney, Arun Rath, and David Krasnow; associate producers Jocelyn Gonzales, Steve Nelson, Michele Siegel, Ave Carrillo, Derek John, Jenny Lawton, and Pejk Malinowski; as well as the many independent producers who tell stories on the show, including Reese Erlich and Wesley Horner, whose stories are quoted here; and the interns who help the team produce *Studio 360* week after week. You all contribute the vitality and spark that has made the program a lively listen for more than a decade.

Thanks also to Dean Cappello, who guided and prodded the show's progress through the public radio landscape; to Chris Bannon for his advice and for filling in whenever necessary; to Peter Wilderotter and Kelly Frost, who took the long journey home with me on 9/11; and to the many people whose business savvy, fund-raising, and marketing have helped keep *Studio 360* on the air, including Leslie Wolfe, Ellen Widmark, Julia Yager, Mark Kausch, and Melissa Barkley at PRI; Alex Villari, Amy Busam, Michele Rusnak, Mitch Heskel, Betsy Gardella, Phil Redo, Leslie Mardenborough, Ellen Reynolds, and Abigail Feder-Kane at WNYC; and to Barbara Sieck Taylor and David Candow.

A special thank-you for the support of Alisa Miller and Steve Salyer at PRI, and to the boards of directors at PRI and WNYC, especially Deedie Rose, Judy Rubin, and Nicki Tanner. Thanks to the many public radio stations that broadcast the program and the foundations who have generously supported the show in its first ten years, including The Corporation for Public Broadcasting, The National Endowment for the Arts, The National Endowment for the Humanities, and the Sloan Foundation. A big thank-you to *Studio 360* listeners all across the country and the world—we created the program for you!

The closest creative circle is here at home. I could not have written this book without the clear-eyed and loving support of my mother, Janet Handler Burstein, who was the best first reader I could have asked for; the long walks with Kate McCaffrey and Ramsey; the sustaining and thought-provoking conversations with Dick Nodell; the friendship of my radio colleagues Sara Fishko, Mary Beth Kirchner, Marge Ostroushko, Melissa Eagan, and Indira Etwaroo; the diversion of sitting in for Leonard Lopate and working with him and his team; the care of Elaine Stern, Jason Neff, Sharon Nathan, Patricia Videgain, Susan Cantor, Diane Thomson, Mark Morris, and Amy Joy Small; the creative release of playing with clay at the NJCVA with Tom Neugebauer and Kirsten Sejda; Shira Nayman for her enthusiasm and excellent questions; Lydia Dean Pilcher for her generosity; Pavla Richterova Perry for her playfulness with my children and her extraordinary photographs; and the support of friends, neighbors, and relations, especially Harold and Lynne Handler, Ethel and Seymour Sobel, Ruth Manos, Ellen Gerstell, Lisa Blitt, Jack

Sobel, Chris Manos, Zoe Manos, Rachel Basch, Julia Doern, Sarah McNamara, Mark and Jessica Zitter, Elaine Charnov, John Barth, Nancy Rosenberg, Sarah McNamara, Diane Jacobs, and Talya and Stu Rothenberg.

I'm thankful, as well, for the love and playfulness of some who are no longer here: my father, Robert Burstein, who shared with me his love of Ella Fitzgerald and Camille Pissarro and applauded every radio program; Dorothy and Bill Maben; and my friend Beth Sobel, with whom I first made music, art, and all sorts of creations.

I am tremendously grateful to Mark Burstein and David Calle for their constant love and support; and to my sons, Ezekiel and Micah Maben, for their lively enthusiasm, curiosity, and love. Most important, my thanks to Mark Maben, for your love and for taking care of so much when *Studio 360* and *Spark* consumed me; without you this would have been impossible.

Finally, my great thanks to the artists who have graced *Studio 360* with their stories and have graciously allowed me to share them in this book. I have learned so much from all of you about how creativity works.

Notes on Sources

CHAPTER 1: ENGAGING ADVERSITY

5 Link to image of Chuck Close's *Big Self-Portrait* (1968) in the collection of the Walker Arts Center: http://collections.walkerart.org/item/object/77

6 Details of Chuck Close's early life: Jon Marmor, "Close Call," *University of Washington Alumni Magazine*, June 1997, http://www.washington.edu/alumni/columns/june97/close1.html

8 **"I've had face blindness, or prosopagnosia":** Michelle Kung, "Chuck Close Gets Perceptional," *WSJ.com*, June 12, 2009: http://blogs.wsj.com/speakeasy/2009/06/12/chuck-close-gets-perceptional/

9 Link to image of Chuck Close's *Robert/104,072* in the Museum of Modern Art: http://www.moma.org/collection/browse_results.php?criteria=O%3AAD%3AE%3A1156&page_number=3&template_id=1&sort_order=1

10 Link to image of Chuck Close's *Lucas* in the Metropolitan Museum of Art: http://www.metmuseum.org/works_of_art/collection_database/modern_art/lucas_chuck_close/objectview_zoom.aspx?page=228&sort=0&sortdir=asc&keyword=&fp=1&dd1=21&dd2=0&vw=1&collID=21&OID=210004973&vT=1

11 **"work became such a joy":** Paula Span: "Chuck Close, the Big Picture: The Artist's New Perspective After the 'Event' of a Lifetime," *Washington Post*, February 22, 1998.

11 **"There are things I miss":** Ibid.

11 Link to image of Chuck Close's 1997 *Self-Portrait*: http://www.moma.org/collection/browse_results.php?criteria=O%3AAD%3AE%3A1156&page_number=23&template_id=1&sort_order=1

12 Link to image of Chuck Close's *Self-Portrait/Scribble/Etching* (2000): http://www.chuckclose.coe.uh.edu/life/scribble.html

13 Link to image of Chuck Close's tapestry *Brad*: http://www.magnoliaeditions.com/Content/Close/F00001.html

15 **show about memorials on the first anniversary of 9/11:** *Studio 360* program on Memorials with Donald Hall, September 7, 2002: http://www.studio360.org/yore/show090702.html

16 **"They pushed the IV pump":** Donald Hall, *Without* (Boston: Houghton Mifflin, 1998), p. 1.

16 **"looked like a huge condom.":** Ibid., p. 22.

18 **"a week after you died":** Ibid., p. 49.

18 **"Ordinary days were best":** Ibid., p. 51.

18 **"you've returned / before me, bags of groceries upright":** Ibid., p. 52.

19 **"works with the presence and practice of sixty years":** Donald Hall, *The Painted Bed* (Boston: Houghton Mifflin, 2002), p. 60.

20 "poet reading it on *Studio 360*": http://www.studio360.org/yore/show090702.html

20 **"I let her garden go":** *The Painted Bed*, p. 43.

22 Link to Jill Sobule's TED performance: http://www.ted.com/talks/jill_sobule_sings_to_al_gore.html

22 Jill Sobule singing *A Good Life:* http://www.jillsobule.com/resources/GoodLife.mp3

22 **At six she was playing the drums** . . . Stephen Elliot: The Rumpus Interview with Jill Sobule, July 30, 2009, http://therumpus.net/2009/07/the-rumpus-interview-with-jill-sobule/

23 Video of Jill Sobule singing *I Kissed a Girl* http://www.youtube.com/watch?v=k4r41vPTF8k

23 "**I have never made a cent off a record in my life** . . ." Denise Quan: Sponsor Jill Sobule's album, get a spot on it, *CNN.com,* March 24, 2009 http://www.cnn.com/2009/SHOWBIZ/Music/03/24/jill.sobule.album/

23 ibid.

26 Jill Sobule sings "Palm Springs": http://www.jillsobule.com/newcd

CHAPTER 2: MODERN ALCHEMY

27 Link to image of Michael Heizer's *North, East, South, West*: http://www.diabeacon.org/exhibitions/main/83

27 Link to image of Robert Smithson's *Map of Broken Glass*: http://www.diabeacon.org/exhibitions/main/97

27 Link to image of John Chamberlain's *Privet*: http://www.diabeacon.org/exhibitions/main/79

28 Link to image of Joseph Beuys's felt sculptures: http://www.diabeacon.org/exhibitions/main/75

28 Link to image of Richard Serra's *Torqued Ellipses*: http://www.diabeacon.org/exhibitions/main/96

39 Link to video of dancers with Plexiglas wall, *Streb vs Gravity*: http://www.streb.org/V2/company/video.html

40 Link to video of STREB dancers with large wheel: http://www.youtube.com/watch?v=MuTuyvquXls

40 **a potential for some "guillotine action.":** "Ensemble": http://www.streb.org/V2/company/video.html

44 Link to video of installation of Richard Serra's sculptures in the garden at the Museum of Modern Art: http://www.moma.org/interactives/exhibitions/2007/serra/flash.html, "Installation of Sculpture Garden."

45 Link to video of installation of Richard Serra's sculptures in MoMA galleries: http://www.moma.org/interactives/exhibitions/2007/serra/flash.html, "Installation of Second Floor."

45 **"Every piece is balanced,":** Ibid.

46 Link to image of Velásquez's *Las Meninas*: http://www.museodelprado.es/en/the-collection/online-gallery/on-line-gallery/zoom/1/obra/the-family-of-felipe-iv-or-las-meninas/oimg/0/

47 **"When you are designing in brick,":** Richard Weston: *Materials, Form and Architecture,* (New Haven, Yale University Press, 2003), p. 93. http://books.google.com/books?id=O4Ol2T9K5pcC&pg=PT102&dq=Kahn+brick&hl=en&ei=jG9QTKXmGcOC8gbhtuGgAQ&sa=X&oi=book_result&ct=result&resnum=9&ved=0CE8Q6AEwCA#v=onepage&q=Kahn%20brick&f=false

48 **"to roll, to curve, to crease,":** http://www.ubu.com/concept/serra_verb.html

48 Link to image of Richard Serra's *To Lift*: http://www.flickr.com/photos/wallyg/2381934666/

49 Link to Richard Serra's short film *Hand Catching Lead:* http://www.youtube.com/watch?v=_NBSuQLVpK4

50 Link to image of Richard Serra's *One Ton Prop:* http://www.moma.org/collection/browse_results.php?object_id=81294

50 Link to image of Richard Serra's *5:30*: http://www.moma.org/interactives/exhibitions/2007/serra/flash.html

51 Link to image of Picasso's *Chicago Sculpture*: http://www.chicagotribune.com/news/politics/chi-chicagodays-picasso-story,0,1344585.story

52 **"I was harassed, ridiculed, disgraced,":** Richard Serra, *Writings/Interviews* (Chicago: University of Chicago Press, 1994), p. 116.

52 **"anywhere where I could find support.":** Ibid., p. 116.

52 **"Rigging is a dangerous business.":** Ibid., p. 191.

CHAPTER 3: THE CULTIVATED AND THE WILD

58 **"partners in this land":** The poem "Snakes of September," Stanley Kunitz with Genine Lentine: *The Wild Braid: A Poet Reflects on a Century in the Garden* (New York: W.W. Norton, 2005), p. 56.

59 **"At my touch the wild / braid of creation / trembles.":** Ibid., p. 55.

59 **"One of my principles is":** Ibid., p. 79.

59 Link to interview with Stanley Kunitz from a story produced for *Studio 360* by Wesley Horner: http://www.studio360.org/spark

60 **"conceived of the garden as a poem in stanzas.":** Kunitz, *The Wild Braid*, p. 72.

61 Link to Kunitz reading "The Snakes of September" on *Studio 360*: http://www.studio360.org/spark

61 **"dangling head-down, entwined":** Kunitz, *The Wild Braid*, p. 55.

61 **"As with the making of a poem,":** Ibid., p. 57.

62 **"My encounter with this family of owls,":** Ibid., p. 31.

62 **"to hear so clear":** Ibid., p. 107.

63 **"where the bees sank sugar wells":** Ibid., p. 32.

63 **"cut away the heart of a poem":** Ibid., p. 57.

63 **"If the terrain were familiar,":** Ibid., p. 89.

64 Link to image of Stanley Kunitz and Elise Asher's gravestone: http://georgefitzgerald.blogspot.com/2009/03/stanley-kunitz-provincetown-cemetery.html

64 **"he loved the earth so much":** Ibid.

65 Link to image of Mel Chin's *Revival Field*: http://collections.walkerart.org/item/object/7577

70 Link to image of the landscape surrounding Ford's River Rouge plant: http://www.thehenryford.org/rouge/environmentalTours.aspx

71 Link to images of Urban Outfitters headquarters in Philadelphia: http://www.dirtstudio.com/projects_view_project.php?project_id=155808

71 Link to plans for Antioch, Illinois, community park: http://www.dirtstudio.com/projects_view_project.php?project_id=272664

71 **"transforming River Rouge into the model of twenty-first-century sustainable manufacturing.":** Alex Steffen, "River Rouge and Neobiological Industry," Worldchanging archives, October 12, 2004: http://www.worldchanging.com/archives/001387.html

73 Link to images of Vintondale project: http://www.dirtstudio.com/projects_view_project.php?project_id=19476

73 Link to photograph of the remains of the colliery buildings: http://www.flickr.com/photos/pruned/2284113622/in/set–72157603967028641/

76 Information about Julie Bargmann and her work, and photograph of Bargmann playing in the dirt at age four: Lee Graves: "Queen of Slag," *The*

University of Virginia Magazine, summer, 2006, http://www.uvamagazine.org/features/article/queen_of_slag/

CHAPTER 4: GOING HOME

79 Link to *Studio 360* story: "Destination: Omaha," produced by Derek John; host Kurt Andersen: http://www.studio360.org/spark

82 Link to image of William Christenberry's "Red Building in Forest": http://www.pacemacgill.com/williamchristenberry-8-2.html

83 Link to image of William Christenberry's "Red Building in Forest": http://www.corkingallery.com/?q=node/106/album&g2_view=core.ShowItem&g2_itemId=4712

83 Link to image of William Christenberry's "Pear Tree with Storm Cloud": http://www.pacemacgill.com/williamchristenberry-2-1.html

83 Link to image of William Christenberry's "Kudzu with Red Soil": http://www.moma.org/collection/browse_results.php?criteria=O%3AAD%3AE%3A1112&page_number=4&template_id=1&sort_order=1

83 Link to image of William Christenberry's "31 Cent Gasoline Sign": http://www.pacemacgill.com/williamchristenberry-27-5.html

84 Link to image of William Christenberry's "Coleman's Café": http://americanart.si.edu/images/1986/1986.69.13_1b.jpg

87 Link to image of William Christenberry's "China Grove Church": http://americanart.si.edu/collections/search/artwork/?id=4758

87 Link to image of William Christenberry's "Church, Sprott, Alabama": http://americanart.si.edu/images/1984/1984.26.1_1b.jpg

87 Link to image of William Christenberry's "Green Warehouse": http://americanart.si.edu/images/1986/1986.68.5_1b.jpg

87 Link to image of William Christenberry's "Kudzu Devouring Building": http://www.corkingallery.com/?q=node/106/album&g2_view=core.ShowItem&g2_itemId=4706

87 Link to image of William Christenberry's "House and Car": http://www.hemphillfinearts.com/ARTISTS/WAC_j.html

87 Link to image of William Christenberry's *River House*: http://americanart.si.edu/images/1994/1994.92_1b.jpg

87 Link to video of William Christenberry in his studio: http://www.artbabble.org/video/meet-william-christenberry

88 **"When I got to the top of the steps,":** Robert Hirsch, "The Muse of Place and Time: An Interview with William Christenberry," *Afterimage*, November 1, 2005.

88 Link to image of William Christenberry's *KKK Doll*: http://americanart.si.edu/collections/search/artwork/?id=4769

88 **"The young lady at the cash register":** Hirsch, "The Muse of Place and Time."

90 **"not by the banality of evil,":** Teresa Wiltz, "A 'Klan Room' Filled with Relics, but Empty of Import," *Washington Post*, February 19, 2008.

90 **"Only later did I realize":** Chris Waddington, "Old South—Modern Art," *Times-Picayune* (New Orleans), February 14, 1997.

98 **"sit on these sofas, polish this slippery-smooth floor,":** Chimamanda Ngozi Adichie, *Half of a Yellow Sun* (New York: Anchor Books, 2007), p. 6.

98 **"The floors were always very cool,":** "A Novelist Remembers the Desks—Including Her Father's—Where She Learned to Write," *Washington Post*, June 17, 2007.

103 **"Olanna looked at it and could not imagine":** Adichie, *Half of a Yellow Sun*, p. 409.

103 **"took long walks on campus,":** Ibid., p. 539.

CHAPTER 5: IMAGINATION'S WELLSPRING

109 **"I got hooked into cinema verité":** Mira Nair interviewed by John Lithgow: http://athome.harvard.edu/programs/cmn/cmn_video/cmn_2.html

110 **"inspired by the question I had in my mind":** Ibid.

111 **"They're listening to the radio in the rain,":** Ibid.

111 **"When you make a thing personal,":** Ibid.

113 David Plowden's website: http://www.davidplowden.com/

113 Link to image of David Plowden's "Great Northern Railway, Extra 3383, Near Willmar, Minnesota (1955)": http://www.davidplowden.com/?cat=12

113 Link to image of David Plowden's "M.V. Britannic, Arriving in New York City, on It's Last Westbound Crossing (December 1960)": http://www.davidplowden.com/?cat=8

113 Link to image of David Plowden's "Copper Mine and Smelter, Globe, Arizona (1978)": http://www.davidplowden.com/?cat=15#slide-2

114 **"confer a kind of immortality:** http://search.barnesandnoble.com/Small-Town-America/David-Plowden/e/9780810938427

117 **"I have been beset,":** http://beinecke.library.yale.edu/digitallibrary/plowden.html

119 Link to *Studio 360* interview with Richard Ford: http://www.studio360.org/episodes/2006/11/10

124 Link to video of Bill Viola's *Ocean Without a Shore:* http://channel.tate.org.uk/media/26506128001

125 **"So I began to really focus on these altars":** Ibid.

CHAPTER 6: MOTHERS AND FATHERS

137 **Milch says that twenty-five "wiseguys":** Video interview with David Milch at MIT, April 20, 2006: http://mitworld.mit.edu/video/383

139 **"Both as a director and an actor,":** Julian Roman, "Kevin Bacon Interview: Pulling Double Duty as Actor and Director in *Loverboy*," *Movieweb*, June 14, 2006: http://www.movieweb.com/news/NEpSDytv7V1Atp

148 **"a very talented young man.":** R. J. DeLuke, "Dewey Redman: The Sound of a Giant," *All About Jazz*: http://www.allaboutjazz.com/php/article_print.php?id=937

149 **"I walked around the corner and got an espresso,":** http://www.joshuaredman.com/releases.php?num=57

CHAPTER 7: CREATIVE PARTNERS

156 **"The minute I met Robert":** Jonathan Takiff, "Plant and Krauss, an Unlikely Pairing, Roll Into Philly," *Philadelphia Daily News*, July 11, 2008.

157 Link to video for "Gone, Gone, Gone": http://www.robertplantalisonkrauss.com/site.php?content=video

159 **"There's so much life and experience":** Richard Cromelin, "O Brother, Who Would Have Guessed This Pairing?" *Los Angeles Times*, November 11, 2007.

159 **T-Bone Burnett described the rhythms they played:** Jon Pareles, "When It Takes Three People to Make a Duet," *New York Times*, October 21, 2007.

161 **"He came into the office," Schamus recalled:** Steven Frazier, Interview with James Schamus, *CNN Presents*, July 1, 2001: http://transcripts.cnn.com/TRANSCRIPTS/0107/01/cp.00.html

163 **"We're useless! This is such a shoddy disgrace!":** Ang Lee, *The Wedding Banquet*, behind-the-scenes video, MGM, 1993.

170 **"Everyone always tells young writers":** James Schamus and Ang Lee, *Eat Drink Man Woman*, behind-the-scenes video, World Films, 1994.

170 **"It must be something universal":** Ibid.

175 Link to photographs of Robert Venturi's house for his mother: http://www.vsba.com/projects/fla_archive/010slide9.html

176 **Scott Brown was born Denise Lakofski:** For details of Denise Scott Brown's early life: David B. Brownlee, David G. DeLong, and Kathryn B. Hiesinger, *Out of the Ordinary: Robert Venturi, Denise Scott Brown and Associates* (Philadelphia: Philadelphia Museum of Art, 2001), p. 5.

178 **what the team called a "decorated shed,":** Robert Venturi, Denise Scott Brown, and Steven Izenour: *Learning from Las Vegas* (Boston: Massachusetts Institute of Technology, 1972), p. 12.

179 **that Denise Scott Brown "is so intertwined":** Ibid., p. xii.

182 Link to website for James Venturi's film about his parents: http://www.bobanddenise.org/

CHAPTER 8: REWEAVING A SHATTERED WORLD

191 **"You can rebuild this," he says:** *Studio 360* story on Terence Blanchard by Reese Erlich: http://www.studio360.org/spark

192 **"I was crying when I took her to Target":** Siddhartha Mitter, "He Works to Raise Hope and Homes in New Orleans," *Boston Globe*, August 24, 2007.

192 **"I've always respected Spike,":** John Wirt, "Blanchard Bringing Requiem Back to New Orleans," *Baton Rouge Advocate*, November 2, 2007.

193 **"I was a little nervous about it.":** Ibid.

194 **"On the one hand, I had to remain a professional":** Ibid.

194 **"When you see those dead bodies":** Ibid.

195 **"was a hard thing to do . . .":** Mitter, "He Works to Raise Hope and Homes in New Orleans."

198 Link to figures on civilian deaths in the Democratic Republic of Congo: http://www.refugeesinternational.org/where-we-work/africa/dr-congo

198 **"More than 1,000 civilians have been killed,":** "DR Congo: Civilian Cost of Military Operation Is Unacceptable," Oxfam news blog, October 13, 2009: http://www.oxfam.org.uk/applications/blogs/pressoffice/?p=7614&v=newsblog

199 **"I thought to myself, 'This play'":** Patrick Pacheco, "Truth and Dare; The Targets Shift, but Lynn Nottage Always Points a Provocative Pen," *Los Angeles Times*, April 19, 2009.

199 Link to photographs of some of the women Lynn Nottage interviewed: http://www.nytimes.com/interactive/2009/02/11/theater/20090211-ruined/index.html

200 Link to Salima's monologue from *Ruined*: http://www.studio360.org/spark

202 **"Ms. Nottage has endowed the frail-looking Sophie,":** Ben Brantley, "War's Terrors, Through a Brothel Window," *New York Times*, February 11, 2009.

203 **"They have never been to Africa.":** Pacheco, "Truth and Dare."

204 Link to Joel Meyerowitz's photo of the World Trade Center at dusk: http://www.joelmeyerowitz.com/photography/aftermath_slide.html; information about Meyerowitz's studio: Nancy Bernhaut, "Joel Meyerowitz at Ariel Meyerowitz," *Art in America*, June 2002.

204 **"I made that picture, which was very quiet,":** Ibarionex R. Perello, "Joel Meyerowitz—AFTERMATH," *Digital Photo Pro* website: http://www.digitalphotopro.com/profiles/joel-meyerowitz-aftermath.html

CHAPTER 9: GETTING TO WORK

220 **Each year on January 7, Allende organizes the desk:** Janet Hawley, "A Woman of Spirit," *The Age*, March 15, 2008.

220 **On the morning of January 8, she meditates:** Ben Naparstek, "Isabel, the Inventor," *Canberra Times*, March 13, 2004.

220 **"I'm just procrastinating," she said.:** Ibid.

221 **"I live like a monk.":** Catherine Elsworth, "Isabel Allende," *Sunday Telegraph Magazine*, March 23, 2008.

223 **. . . Oates embodies a similar devotion to her work.:** Sybil Steinberg, "Prolific Oates," *Publishers Weekly*, September 12, 2004.

224 **"The structural problems I set for myself":** Joyce Carol Oates, "To Invigorate Literary Mind, Start Moving Literary Feet," *New York Times*, July 18, 1999.

232 **"about that sensation of history cracking open.":** Frank Rich, "Caroline, Kennedy and Change," *New York Times*, December 7, 2003.

Grateful acknowledgment is made for permission to reprint the following:

Excerpts from *Without: Poems* by Donald Hall. Copyright © 1998 by Donald Hall. Reprinted by permission of Houghton Mifflin Harcourt Publishing Company. All rights reserved.

Excerpts from *The Painted Bed: Poems* by Donald Hall. Copyright © 2002 by Donald Hall. Reprinted by permission of Houghton Mifflin Publishing Company. All rights reserved.

Excerpts from *We Are on Our Own* by Miriam Katin. Copyright © 2006 by Miriam Katin.

Excerpts from the following: "The Snakes of September." Copyright © 1985 by Stanley Kunitz; "Touch Me." Copyright © 1995 by Stanley Kunitz; "The Testing-Tree." Copyright © 1968 by Stanley Kunitz, from *The Collected Poems* by Stanley Kunitz. Used by permission of W.W. Norton & Company, Inc.

From *The Wild Braid: A Poet Reflects on a Century in the Garden* by Stanley Kunitz and Genine Lentine. Copyright © 2005 by Stanley Kunitz and Genine Lentine. Used by permission of W.W. Norton & Company, Inc.

Excerpts from the *Studio 360* interview with Tony Kushner used by permission of Tony Kushner. All rights reserved.

Excerpts from *Half of a Yellow Sun* by Chimamanda Ngozi Adichie. Copyright © 2006 by Chimamanda Ngozi Adichie. Reprinted by permission of Anchor, a division of Random House, Inc.

Excerpts from *Ruined* by Lynn Nottage. Copyright © 2009 by Lynn Nottage.

About the Author

Julie Burstein is fascinated by the roots of creativity, and she has pursued that passion as a public radio reporter, producer, and host for WNYC, NPR, and PRI. Julie has produced radio series for Carnegie Hall and the New York Philharmonic, and was the first arts reporter for WHYY in Philadelphia. In 2000, PRI and WNYC asked Julie to create *Studio 360* with Kurt Andersen, public radio's Peabody Award–winning guide to creativity, pop culture, and the arts, and she led the show's creative team for nine years. *Spark: How Creativity Works* is her first book. Julie lives in New Jersey with her husband and two sons.

About *Studio 360*

Peabody Award–winning *Studio 360* from Public Radio International and WNYC is public radio's smart and surprising guide to what's happening in pop culture and the arts, hosted by novelist and journalist Kurt Andersen. The show is broadcast on 140 stations across the country, and is heard by 500,000 listeners each week. For more about the program, visit studio360.org.